T0144604

Bioterrorism:
A Guide For Facility Managers

Bioterrorism:
A Guide For Facility Managers

by
Joseph F. Gustin

THE FAIRMONT PRESS, INC.

CRC Press
Taylor & Francis Group

Library of Congress Cataloging-in-Publication Data

Gustin, Joseph F., 1947-
    Bioterrorism: a guide for facility managers / by Joseph F. Gustin.
        p. cm.
    Includes bibliographical references and index.
    ISBN 0-88173-468-3 (print) -- ISBN 0-88173-469-1 (electronic)
        1. Buildings--Security measures. 2. Security systems. 3.
Buildings--Environmental engineering. 4. Bioterrorism--Prevention. I. Title.
    TH9705.G87 2005
    658.4'77--dc22

                                                    2005040898

Published by The Fairmont Press, Inc.
700 Indian Trail
Lilburn, GA 30047
tel: 770-925-9388; fax: 770-381-9865
http://www.fairmontpress.com

Distributed by Taylor & Francis Ltd.
6000 Broken Sound Parkway NW, Suite 300
Boca Raton, FL 33487, USA
E-mail: orders@crcpress.com

Distributed by Taylor & Francis Ltd.
23-25 Blades Court
Deodar Road
London SW15 2NU, UK
E-mail: uk.tandf@thomsonpublishingservices.co.uk

Printed in the United States of America
10 9 8 7 6 5 4 3 2 1

0-88173-468-3 (The Fairmont Press, Inc.)
0-8247-2158-6 (Taylor & Francis Ltd.)

Table of Contents

Preface

While the art of facility management lies in integrating people with their physical environment, the science of facility management lies in executing the complexities of this integration.

Facility management, in its traditional form, involves responsibility for the care and maintenance of a site's physical property. These "bricks and mortar" issues—HVAC, lighting, electric, plumbing, space allocation, and grounds maintenance continue to be the mainstay of facility management.

However, in its hybrid form, the processes involved in addressing these "mainstays" of facility management have become more complex. With the proliferation of regulatory mandates, at the federal, state and local levels, worker compensation issues, increased employee litigation, and workplace violence, the role of the facility manager has been redefined. Adding to these complexities is a new and emergent factor—the insidious threat of terrorism. And, whether that threat is biological, chemical or radiological in nature, facility managers must be attuned to their role in ensuring the safety of building occupants and the protection of the physical site, itself.

Bioterrorism: A Guide for Facility Managers addresses these issues. It provides a macro perspective on the issues of biological, chemical and radiological terrorism. It defines the various threats to a facility and outlines the various strategies to reduce occupant and building vulnerability.

Introduction

Facility managers are the professionals most responsible for integrating people with their physical environment. As such, facility management is both a people and an environmental issue. The hybrid role of the facility manager as operations manager and compliance officer involves people and productivity, and the costs of managing each. The facility manager must coordinate policies and operations with industry standards and practices, as well as with regulatory mandates.

Bioterrorism: A Guide for Facility Managers provides a rationale for systematically identifying and evaluating the key areas of practice management. For example, Chapter 1 lays the groundwork for gaining an understanding of terrorism and the differences in terrorist threats. It also discusses the risks that terrorism presents to companies, building owners and facility managers.

Chapter 2 reviews building vulnerability and the elements needed to design, create and maintain compliant work environments. From conducting baseline assessment to discussing ventilation and filtration issues, specific recommendations are provided to facility managers for addressing the key elements in building systems design and maintenance.

Chapter 3 addresses HVAC systems and how the risk of biological, chemical and radiological threats impact systems operations and maintenance.

Chapter 4 presents an in-depth discussion on safeguarding buildings. For example, facts about airborne hazards are presented, along with both immediate and long-term actions that can be taken to mitigate the effects of biological, chemical and radiological releases.

Chapter 5 discusses the issue of RDDs, or "dirty bombs," and what facility managers can do following an explosion, should it occur.

Bioterrorism: A Guide for Facility Managers is unique in its scope. It focuses upon the awareness of terrorist threat. It also focuses upon practice management. By doing so, it turns the challenges of facility management into opportunities for the facility manager. These opportunities are manifested in an enhanced productivity that aligns itself with ensuring the safety of building employees, occupants and tenants, as well as with business operations.

Chapter 1

Chemical, Biological and Radiological Weapons and the Facility Manager— An Overview

INTRODUCTION

T ERRORISM, as defined by the United States Federal Bureau of Investigation (FBI), is the "unlawful use of force or violence committed by a group or individual against persons or property to intimidate or coerce a government, the civilian populations, or any segment thereof, in furtherance of political or social objectives."

In recent years, the world has witnessed countless acts of terrorism. Over the past decade alone, numerous and wanton terrorist acts have occurred. And the United States has not been exempt. The most notable examples of terrorism include the first bombing of the World Trade Center in 1993, the bombing of the Murrah Federal Building in Oklahoma City in 1995, and the September 11, 2001 attacks. Following the September 11th attacks, terrorist activities took a new turn as evidenced by the sending of anthrax-laced mail to various locations and targets. U.S. interests were being attacked with a "new" and more insidious weapon.

GOALS

Because of its goals, a terrorist attack can occur with or without warning. The goal or purpose of an attack is to cause disruption of routine, economic loss, loss and/or disruption of critical resources and vital services, loss of life and emotional devastation. Certainly, in each of the cases listed above, the terrorist goals were met.

WEAPONS CLASSIFICATION

Generally speaking, terrorist weapons can be classified into the following basic types:

- Explosive
- Incendiary
- Biological
- Chemical
- Nuclear

Explosives
According to the United States Department of Transportation (DOT), an explosive can fall into either of the following definitions:

- A substance or article, including a device that is designed to explode (e.g., an extremely rapid release of gas and heat); or

- Any substance or article, including a device, which by chemical reaction within itself, can function in a similar manner even if the substance or article in question is not designed to explode.

Incendiaries
Incendiary devices are any mechanical, electrical or chemical devices that are used intentionally to ignite combustion and start a fire. Whether the devices are mechanical, electrical or chemical, they are comprised of three basic components—a fuse/igniter; the body of the device (e.g., container); and the filler. Depending on its intended use the body, or container, is made of glass, metal, plastic or paper. If the device contains chemicals, it will be usually be made of metal, or some other non-breakable material. A liquid accelerant will usually be placed in a breakable container such as glass.

Biological Weapons
Humans are exposed daily to biological agents. Natural resistance to these agents, inoculations, good hygiene and nutrition help ward off any ill-effects or harm that these biological agents may pose. However, since these biological agents are found in nature, they are easily accessible, have the potential for rapid spread and can be contagious.

Several biological agents can be adapted and used as terrorist weapons. These include anthrax which is found in hoofed animals, tularemia (rabbit fever), and the pneumonic plague, which is sometimes found in prairie dog colonies, and botulism, which is found on food products. According to the United States Department of Health and Human Services Centers for Disease Control (CDC), biological agents of the greatest concern are:

- anthrax
- smallpox
- plague
- tularemia
- botulism, and
- viral hemorrhagic fevers.

These agents can be easily transmitted from person to person. As such, they can result in high mortality rates and have an impact on public health.

Routes of Exposure
The way biological agents make their way into a human's system is through inhalation, ingestion and skin absorption/contact. Exposure to a biological incident may not be detected readily because the onset of some symptoms may take days to weeks. Since biological agents are usually odorless and colorless, typically there will be no characteristic symptoms to signal an incident. Therefore, the number of victims and the geographical areas affected may be greater than the initial area. Because the infected persons are not aware of their exposure they travel to their homes, infecting other people. On the other hand, some effects may be very rapid, producing symptoms in as little as four to six hours.

Differences Between Agents
Chemical, biological and radiological material, as well as industrial agents, can be introduced into humans through the following ways:

- In the air that is breathed
- The water that is drunk, or
- Physical contact on surfaces

These chemical, biological and radiological materials can be dispersed in a number of ways ranging from simple to complex methods. These methods include placing a container in a heavily used area; opening that container; using conventional garden/commercial spray devices to disperse the contents of that container; or detonating an improvised explosive device.

Chemical incidents are characterized by the rapid onset of medical symptoms—minutes to hours—and easily observed signatures. These easily observable signatures include: colored residue, dead foliage, pungent odor, and dead insect and animal life. See Table 1-1.

Biological incidents, however, are slower to detect. The onset of symptoms requires days to weeks and typically, there will be no characteristic signatures. Also, the affected area may be greater because contaminated individuals—not aware of their contamination—may move beyond the affected area. This movement beyond an affected area has implications for infecting a greater number of people or a wider area. See Table 1-2.

Radiological incidents also are slower to detect. Days to weeks are required for the onset of symptoms and again, as with the case of biological incidents, there are typically no characteristic signatures. Radiological materials are colorless and odorless and therefore, are not recognizable by the senses. And, with the case of biological incidents, contaminated individuals, not aware of their contamination, may move beyond the affected areas, thereby affecting a greater number of people, or a wider geographic area. See Table 1-3.

Chemical Agents

Chemical agents that might by used by terrorists range from warfare agents to toxic chemicals commonly used in industry. The CDC identifies the following nerve agents that pose a threat as chemical weapons:

- VX
- Sarin
- Tabun
- Soman
- Vesicants or blistering agents such as sulfur mustard
- Toxic chemicals such as:
  — Cyanide, and
  — Ricin

Table 1-1. Indicators of a Possible Chemical Incident

Dead animals/birds/fish	Numerous animals-wild/domestic, small/large-birds and fish in the same area.
Lack of insect life	If normal insect activity-ground, air, and/or water-is missing, then check the ground/water surface/shoreline for dead insects. If near water, check for dead fish/aquatic birds.
Physical symptoms	Numerous individuals experiencing unexplained water-like blisters, wales (like bee stings), pinpointed pupils, choking, respiratory ailments and/or rashes.
Mass casualties	Numerous individuals exhibiting unexplained serious health problems ranging from nausea to disorientation to difficulty in breathing to convulsions and death.
Definite pattern of casualties	Casualties distributed in a pattern that may be associated with possible agent dissemination methods.
Illness associated with confined geographic area	Lower attack rates for people working indoors versus outdoors, or outdoors versus indoor.
Unusual liquid droplets	Numerous surfaces exhibit oily droplets/film; numerous water surfaces have an oily film-with no recent rain.
Areas that look different in appearance	Not just a patch of dead weeds, but trees, shrubs, bushes, food crops, and/or lawns that are dead, discolored, or withered—with no current drought.
Unexplained odors	Smells may range from fruity to flowery to sharp/pungent to garlic/horseradish-like to bitter almonds/peach kernels to new mown hay. It is important to note that the particular odor is completely out of character with its surroundings.
Low-lying clouds	Low-lying cloud/fog-like condition that is not explained by its surroundings.
Unusual metal debris	Unexplained bomb/munitions-like material, especially if it contains a liquid-with no recent rain.

Source: United States Central Intelligence Agency.

Table 1-2. Indicators of a Possible Biological Incident

Unusual numbers of sick or dying people or animals	As a first responder, strong consideration should be given to calling local hospitals to see if additional casualties people or animals with similar symptoms have been observed. Casualties may occur hours to days or weeks after an incident has occurred. The time required before symptoms are observed is dependent on the radioactive material used and the dose received. Additional symptoms likely to occur include unexplained gastrointestinal illnesses and upper respiratory problems similar to flu / colds.
Unscheduled and unusual spray being disseminated	Especially if outdoors during periods of darkness.
Abandoned spray devices	Devices will have no distinct odors.

Source: United States Central Intelligence Agency.

Criteria for determining priority chemical agents include:

- Chemical agents already known to be used as weaponry;
- Availability of chemical agents to potential terrorists;
- Chemical agents likely to cause major morbidity or mortality;
- Potential of agents for causing public panic and social disruption; and
- Agents that require special action for public health preparedness.

Other categories of chemical agents include:
- Nerve agents
  — Tabun
  — Sarin
  — Soman
  — GF
  — VX
- Blood agents
  — Hydrogen cyanide
  — Cyanogens chloride

Table 1-3. Indicators of a Possible Radiological Incident

Unusual numbers, of sick or dying people or animals	As a first responder, strong consideration should be given to calling local hospitals to see if additional casualties with similar symptoms have been observed. Casualties may occur hours to days or weeks after an incident has occurred. The time required before symptoms are observed is dependent on the radio-active material used and the dose received. Additional symptoms include skin reddening and, in severe cases, vomiting.
Unusual metal debris	Unexplained bomb/munitions-like material.
Radiation symbols	Containers may display a radiation symbol.
Heat emitting material	Material that seems to emit heat without any sign of an external heating source.
Glowing material/particles	If the material is strongly radioactive, then it may emit a radio luminescence.

Source: United States Central Intelligence Agency.

- Blister agents
  — Lewisite
  — Nitrogen and sulfur mustards
  — Cyanogen chloride
- Heavy metals
  — Arsenic
  — Lead
  — Mercury
- Volatile toxins
  — Benzene
  — Chloroform
  — Trihalomethanes
- Pulmonary agents
  — Phosgene
  — Chlorine
  — Vinyl chloride
- Incapacitating agents;

— BZ
• Pesticides: persistent and non-persistent
• Dioxins, furans and polychlorinated biphenyls (PCBs)
Chemical agents fall into the following classes:
• Nerve agents
• Blister agents
• Blood agents
• Choking agents
• Irritants

Nerve agents disrupt nerve impulse transmissions. The victims of a nerve agent attack will experience uncontrolled salivation, lacrimation (tears), muscle twitching and contraction. Nerve agents resemble a heavy oily substance. Aerosols are the most efficient distribution. Small explosions and spray device equipment to generate mists may be used. Nerve agents kill insect life, birds and other animals, as well as human beings. Many dead animals at the scene of an incident may be an outward warning sign or detection clue that a nerve agent attack has occurred.

Blister agents, also called vesicants, cause redness and is possibly followed by blisters. They are similar in nature to other corrosive materials. Blister agents readily penetrate layers of clothing and are quickly absorbed into the skin.

These agents are heavy, oily liquids dispersed by aerosol or vaporization. As such, small explosions may occur or spray equipment may be used to disperse the agents. In a pure state, they are nearly colorless and odorless. Slight impurities, however, give them a dark color and an odor suggesting mustard, garlic or onions. Outward signs of blister agents include complaints of eye and respiratory irritation along with reports of a garlic-like odor.

Blood agents interfere with the ability of blood to transport oxygen. The ultimate result is asphyxiation. Common blood agents include hydrogen cyanide (AC) and cyanogens chloride (CK). Cyanide and cyanide compounds are common industrial chemicals. CK can cause tearing of the eyes and irritate the lungs. All blood agents are toxic at high concentrations and lead to rapid death. Affected persons require removal to fresh air and respiratory therapy.

Under pressure, blood agents are liquids. In pure form, they are gases. All forms have the aroma of burned almond or peach kernel. They are common industrial chemicals and are readily available.

Choking agents cause stress to respiratory system tissues. Severe distress causes edema (fluid in the lungs) which can result in asphyxiation. This type of asphyxiation resembles drowning. Chlorine (CL) and phosgene (CG), common industrial chemicals, are choking agents. Clinical symptoms of chlorine and phosgene include severe eye irritation and respiratory distress (coughing and choking). The choking agent chlorine has the easily recognizable odor of chlorine. Phosgene has the odor of newly cut hay. Since both are gases, they must be stored and transported in bottle or cylinders.

Irritants cause respiratory distress and tearing. They are designed to incapacitate. They can cause intense pain to the skin, especially in moist areas of the body. Irritants are also called riot control agents or tear gas. Generally speaking, they are not lethal. However, under certain circumstances, they can result in asphyxiation. Common irritants are Mace® (CN), tear gas (CS) and capsicum/pepper spray. Table 1-4 is a quick reference that describes the biological agents, their characteristics and their implications. Table 1-5 is a quick reference that describes the chemical agents, their characteristics and their implications

Nuclear Weapons

Humans are exposed daily to low levels of radiological substances—sun, soil, X-rays. It is the exposure to an uncontrolled and massive dose of radioactive material that threatens life and well-being. Nuclear material, used in conjunction with explosives, is called a dirty bomb. The possible sources of radiation are the nuclear bomb, a dirty bomb, or a material release.

There are three types of radiation emitted from nuclear material:
* Alpha
* Beta
* Gamma

Alpha particles are the heaviest and most highly charged of the nuclear particles. Alpha particles, however, cannot travel more than a few inches in air. They are completely stopped by an ordinary sheet of paper or the outermost layer of dead skin that covers the body. However, if ingested through eating, drinking, or breathing, they become internal hazards and cause massive internal damage.

Beta particles are smaller and travel much faster than alpha particles. Typical beta particles can travel several millimeters through tissue.

Table 1-4. Biological Agent Quick Reference

Biological Agents

AGENT	INCUBATION	LETHALITY	PERSISTENCE	DISSEMINATION
Bacteria				
Anthrax	1-5 days	3-5 days fatal	Very stable	Aerosol
Cholera	12 hours-6 days	Low with treatment High without	Unstable Stable in saltwater	Aerosol Sabotage of water
Plague	1-3 days	1-6 days fatal	Extremely stable	Aerosol
Tularemia	1-10 days	2 weeks moderate	Very stable	Aerosol
Q fever	14-26 days	Weeks?	Stable	Aerosol Sabotage
Viruses				
Smallpox	10-12 days	High	Very stable	Aerosol
Venezuelan Equine Encephalitis	1-6 days	Low	Unstable	Aerosol Vectors
Ebola	4-6 days	7-16 days fatal	Unstable	Aerosol Direct contact
Biological Toxins				
Botulism toxins	Hours to days	High without treatment	Stable	Aerosol Sabotage
Staphylococcal enterotoxin B	1-6 days	Low	Stable	Aerosol Sabotage
Ricin	Hours to days	10-13 days fatal	Stable	Aerosol Sabotage
Tricothecene mycotoxins (T2)	2-4 hours	Moderate	Extremely stable	Aerosol Sabotage

Source: *hld.sbccom.army.mil; http://usamriid.army.mil/education/bluebook.html*

Table 1-5. Chemical Agent Quick Reference
Chemical Agents

AGENT	SIGNS, SYMPTOMS	DECONTAMINATION	PERSISTENCE
Nerve Agents			
Tabun (GA)	Salivation Lacrimation Urination Defecation Gastric disturbances Emesis	Remove contaminated clothing Flush with a soap and water solution for patients Flush with large amounts of 5% bleach and water solution for objects	1-2 days if heavy concentration
Sarin (GB)			1-2 days; will evaporate with water
Soman (GD)			Moderate, 1-2 days
V Agents (VX)			High; lasts 1 week high concentration As volatile as motor oil
Vesicants (Blister Agents)			
Sulfur Mustard (H) Distilled Mustard (HD) Nitrogen Mustard (HN1,3)	Acts first as a cell irritant, then as a cell poison. Conjunctivitis, reddened skin, blisters, nasal irritation, inflammation of throat and lungs.	Remove contaminated clothing; Flush with soap and water solution for patients; Flush with large amounts of a 5% bleach and water solution for objects	Very high, days to weeks
Mustargen (HN2)			Moderate
Lewisite (L)	Immediate pain; blisters later		Days, rapid hydrolysis with humidity
Phosgene Oxime (CX)	Immediate pain; blisters later; necrosis equivalent to second & third degree burns		Low, 2 hours in soil

(Continued)

Table 1-5. (*Continued*)

Chemical Asphyxiants (Blood agents)

Hydrogen Cyanide (AC)	Cherry red skin or 30% cyanosis.	Remove contaminated clothing.	Extremely volatile, 1-2 days
Cyanogen Chloride (CK)	Gasping for air. Seizures prior to death. Effect	Flush with a soap & water solution for people; large amounts of 5% bleach and	Rapidly evaporates & disperses
Arsine(SA)	similar to asphyxiation, but more sudden	water solution for objects.	Low

Source: *hld.sbccom.army.mil; http://usamriid.army.mil/education/bluebook.html*

Generally speaking, they do not penetrate far enough to reach vital organs. External exposure, i.e., outside the body, to beta particles is normally thought of as a slight danger.

Prolonged exposure of the skin to large amounts of beta radiation however, may result in skin burns. Beta-emitting contamination enters the body from eating, drinking, breathing and unprotected open wounds. Like alpha particles, they are primarily internal hazards.

Gamma rays are a type of electromagnetic radiation transmitted through space in the form of waves. These rays are pure energy and are the most penetrating type of radiation. Gamma rays can travel great distances and penetrate most material. Gamma rays can affect all human tissue and organs as they pass through the body. Gamma radiation has instinctive and short-term symptoms such as skin irritation, nausea, vomiting, high fever, hair loss and dermal burns. Acute radiation sickness occurs when an individual is exposed to a large amount of radiation within a short period of time. Materials used to shield gamma rays and X-rays include such thick materials as lead, steel, or concrete.

THE RISKS

As seen from the descriptions of the various weapons, the risks associated with explosive and incendiary devices are more obvious. On the other hand, biological, chemical or nuclear (radiological) agents are less obvious. For example, in terms of a facility's immediate surroundings, the odor of an accelerant used in incendiary devices is obvious. However, the release of mists, liquids or vapor clouds is less obvious. Table 1-6 is a hazards checklist that outlines the types of facility sites and the potential hazards that are present in these facilities. Other examples of less obvious "occurrences" or "events" may include:

- Explosions that seem to destroy only the package, but which may release biological contaminants, radiation or chemicals;

- Unscheduled spraying or abandoned spraying devices, which may indicate a release of biological or chemical contaminants;

- Hazardous materials/lab equipment that are not site specific, indicating a possible biological, nuclear (radiological) or chemical contamination;

- Numerous sick/dead animals, fish, birds, indicating a possible biological or chemical dispersion in the area.

While unattended packages left in high-risk areas may be indicative of incendiary or explosive devices, those untended packages may also contain biological, nuclear or chemical contaminants.

Personal/Physical Signals

The consequences of these less obvious contaminations on the population can be manifested in various ways. For example, chemical contamination can result in persons reporting unusual tastes/odors; persons salivating, tearing or having uncontrolled muscle spasms. Impaired breathing and redness of skin are also indicative of chemical contamination. Mass casualties without obvious trauma, as well as distinct patterns of casualties with common symptoms, can also be the result of chemical contamination. Biological contamination can produce large numbers of people seeking medical attention with symptoms that are uncharacteristic of the season.

In either case, however, knowing what to do when such obvious or less obvious events occur becomes of paramount importance.

IMPLICATIONS FOR THE FACILITY MANAGER

Recent terrorist threats and the use of biological and chemical agents against civilians have exposed our vulnerability to terrorist acts. As such,

Table 1-6. Hazards Checklist Structural Alarms

If these sites are involved:	Look for these hazards:
Hospitals, medical labs or clinics	Biological waste
	Infectious patients or lab animals
	Radioactive materials
	Compressed and anesthetic gases
	Cryogenic liquid oxygen
	Syringes, sharp instruments
	Ethylene oxide
	Nitrous oxides
	Radioisotopes
Manufacturing and processing	Resins (alkyds, vinyls)
	Pigments (dry metallic powders)
	Compressed gases
	Ammonia refrigerant
	Lubricating and cooking oils
	Dusts and cotton fibers
	Aerosols
	Radioisotopes, radioactive materials
Retail and commercial	Flammable gases & liquids
	Corrosives
Business offices	Cleaning products
	Copy chemicals
Residences and hotels	Gasoline & solvents
	Pesticides & fertilizers
	Pool products
	Paint
	Cleaning supplies
	Illegal drug labs
Schools	Pool products
	Laboratory chemicals
Farms	Pesticides

(*Continued*)

Table 1-6. (*Continued***)**

	Oxidizers
	Anhydrous ammonia
Construction	Ammonium nitrate fuel oil mixture
	Compressed gases
	Solvents
	Radioisotopes
Mining	Methane
	Cola dust & metal sulfide ores
	Radioactive materials (uranium)

Source: U.S. Department of Health and Human Services, Centers for Disease Control and Prevention (CDC)

it highlights the need to enhance our capacity to detect and control these acts of terrorism.

In the past, most planning for emergency response to terrorism has been concerned with overt acts such as bombings. Other overt terrorist acts include the use of chemical agents. Since chemical agents are absorbed through inhalation or through the skin or mucous membrane, their effects are usually immediate and obvious. Such overt acts require immediate response.

On the other hand, attacks with biological agents are more likely to be covert. Biological attacks present different challenges and require an additional dimension to the emergency planning process. The effects of biological attacks do not have an immediate impact because of the delay between exposure and onset of illness. As such, the likelihood for a disease to be disseminated through the population by person-to-person contact increases.

Certain chemical agents can also be disseminated through contaminated food or water. The Centers for Disease Control (CDC) reported that the vulnerability of the food supply was illustrated in Belgium, in 1999, when chickens were unintentionally exposed to dioxin-contaminated fat used to make animal feed. Because the contamination was not discovered for months, the dioxin, a cancer-causing chemical that does not cause immediate symptoms in humans, was probably present in chicken meat and eggs sold in Europe in 1999. This dioxin episode demonstrates how a covert act of food-borne biological or chemical terrorism could affect not only human or animal health, but industrial commerce as well. This incident underscores the need for prompt diagnoses of unusual or suspicious health problems in animals as well as in humans. And this lesson was demonstrated by the outbreak in New York City in 1999 of mosquito-borne West Nile virus in birds and humans.

Detection, diagnosis and mitigation of illness and injury caused by biological and chemical terrorism is a complex process. It involves numerous partners and additional activities. Meeting this challenge adds a new dimension to the emergency planning process. As such, building owners and facility managers will need to work closely with their employees and building occupants so that all individuals are aware of potential hazards. Also, building owners and facility managers will need to work in concert with their local police and fire departments, as well as with local health providers and agencies. By doing so, the strategies required to prevent and/or mitigate illness or injury caused by biological and chemical acts can be developed.

Sources

Gustin, Joseph F. *Disaster and Recovery Planning: A Guide for Facility Managers, 3rd ed.*, Lilburn, GA: The Fairmont Press, Inc., 2004.

Gustin, Joseph F. *The Facility Manager's Handbook*, Lilburn, GA: The Fairmont Press, Inc., 2003.

Gustin, Joseph F. *Safety Management: A Guide for Facility Managers*, New York: UpWord Publishing, Inc., 1996.

United States Department of Health and Human Services, Centers for Disease Control and Prevention (CDC), *Biological and Chemical Terrorism: Strategic Plan for Preparedness and Response, Recommendations of the CDC Strategic Planning Workgroup*, April 21, 2000/49(RR04); 1-14.

Chapter 2

Building Vulnerability

Recent terrorist events have demonstrated the vulnerability of U.S. buildings and other tenanted facilities, including workplaces, to such attacks. Whether such attacks are directed at the U.S. Postal Service, Congressional representatives, or high profile media representatives, the terrorist threat of expanding targets remains present.

Because there is no single process or formula that can determine a building's vulnerability to a chemical, biological or radiological attack, the likelihood of a specific building being targeted for such a terrorist attack is difficult to predict. However, there are various procedures, or actions, that facility managers and building owners can take to protect a building's occupants from a chemical, biological or radiological attack. These procedures, or actions, include changing or transforming buildings into less attractive targets. This is accomplished by:

(1)   increasing the difficulty of introducing a chemical, biological or radiological agent;

(2)   increasing the ability to detect terrorists and terrorist activities before those activities are carried out; and

(3)   incorporating procedures and plans that can mitigate the effects of a chemical, biological or radiological release.

It is important to note that while each of these actions is necessary to enhance the safety and protection of a building's occupants, no one building or its occupants can be fully protected from a terrorist determined to release a chemical, biological or radiological agent. However, occupant injuries or fatalities can be minimized by undertaking the actions listed above.

In order to determine how best to implement these actions so that injuries and fatalities can be minimized, facility managers and building owners should perform a baseline assessment of their facilities.

THE BASELINE ASSESSMENT

As previously noted, recent terrorism events have increased concern regarding the vulnerability of tenanted, or occupied buildings to chemical, biological or radiological threats. In terms of the bioterrorist threat, building dynamics and airflow patterns are of particular concern. A building's dynamics and airflow patterns are impacted by the heating, ventilating and air conditioning systems in place. Because these systems can become an entry point, as well as a distribution point for contaminants, including the chemical, biological or radiological agent, facility mangers and building owners need valid and reliable information to

[]   decrease the likelihood, or the effects of an attack; and
[]   respond quickly and appropriately in the event of an incident.

This information can help facility managers and building owners to determine how to modify buildings for better air protection and security, as well as how to design new buildings to be more secure. Additionally, this information should provide the basic information needed during a chemical, biological or radiological incident, should such an incident occur.

The baseline assessment begins by conducting a walk-through inspection of the building or facility. This evaluation includes Heating, Ventilation and Air Conditioning (HVAC), fire protection and life-safety systems. During this evaluation or inspection, the facility manager/building owner must compare the most up-to-date design drawings available to the operation of the current systems. This assistance may require advice and direction from vendors and outside consultants. Without this baseline knowledge, it is difficult to accurately identify what impact a particular security modification may have on building/facility operations. While it is important to understand how the existing building systems function, the systems need not operate per design before implementing security measures. The National Institute of Occupational Safety and Health (NIOSH) recommends the following list of items to consider during the inspection. Please note that this is a partial list:

•   What is the mechanical condition of the equipment?

- What filtration systems are in place? What are their efficiencies?

- Is all equipment appropriately connected and controlled? Are equipment access doors and panels in place and appropriately sealed?

- Are all dampers (outdoor air, return air, bypass, fire and smoke) functioning? Check to see how well they seal when closed.

- How does the HVAC system respond to manual fire alarm, fire detection, or fire-suppression device activation?

- Are all supply and return ducts completely connected to their grilles and registers?

- Are the variable air volume (VAV) boxes functioning?

- How is the HVAC system controlled? How quickly does it respond?

- How is the building zoned? Where are the air handlers for each zone? Is the system designed for smoke control?

- How does air flow through the building? What are the pressure relationships between zones? Which building entryways are positively or negatively pressurized? Is the building connected to other buildings by tunnels or passageways?

- Are utility chases and penetrations, elevator shafts and fire stairs significant airflow pathways?

- Is there obvious air infiltration? Is it localized?

- Does the system provide adequate ventilation given the building's current occupancy and functions?

- Where are the outdoor air louvers? Are they easily observable? Are they or other mechanical equipment accessible to the public?

- Do adjacent structures or landscaping allow access to the building roof?

SPECIFIC RECOMMENDATIONS

According to NIOSH, securing mechanical rooms should be a first priority and all attempts should be made to do so. The specific recommendations for building protection can be divided into four general categories:

1.   Things not to do

2.   Physical security

3.   Ventilation and filtration

4.   Maintenance, administration and training

Things Not To Do
Of prime importance in building protection against a chemical, biological or radiological release is the protection of the building systems and the building occupants. Some efforts to protect the building could have adverse effects on the building's indoor environmental quality. Building owners and facility managers should understand how the systems operate and assess the impact of security measures on those systems. A first step to doing this is to follow the recommendations made by NIOSH. These recommendations begin with the actions not taken. They are listed in the order of priority when securing a building against a possible chemical, biological or radiological attack:

1.   **Do not permanently seal outdoor air intakes**. Buildings require a steady supply of outdoor air appropriate to their occupancy and function. This supply should be maintained during normal building operations. Closing off the outdoor air supply vents will adversely affect the building occupants. The likely result of this action will be a decrease in indoor environmental quality and an increase in indoor environmental quality complaints.

2.   **Do not modify the HVAC system without first understanding the effects on the building systems or the occupants**. This caution directly relates to the recommendation that building owners and managers should understand the operation of their building systems. If there is uncertainty about the effects of a proposed modification, a qualified professional should be consulted.

3. **Do not interfere with fire protection and life safety systems**. These systems provide protection in the event of fire or other types of events. They should not be altered without guidance from a professional specifically qualified in fire protection and life safety systems.

Physical Security

The physical security of a building—its entry, storage, roof and mechanical areas, as well as securing access to the outdoor air intakes of the building's HVAC system, are as important as securing the building's internal systems and occupants. If the outside areas and perimeters of the building are secure, then it becomes harder to attack the building's internal systems and occupants.

The first step in securing the physical security of a building is in assessing its vulnerabilities. Each building/facility varies greatly in its vulnerability to a chemical, biological radiological (CBR) attack, with the multi-national, multi-million dollar corporation being more at risk than a small retail establishment. Therefore, a thoughtful and accurate assessment specific to each building can effectively address its physical security. The simple act of locking the mechanical room doors is a low-cost measure than can be implemented in most buildings. Locking the mechanical room doors also will not substantially inconvenience the building users.

Costlier measures would include such items as increased security personnel, package x-ray equipment, etc.. Once a building's assets, characteristics and use has been determined to be a potential or likely terrorist target, the facility manager and building owner can determine the protection level for their facility.

Security Measures

Security measures applicable to many building types include the following:

1. **Prevent access to outdoor air intakes**. One of the most important steps in protecting a building's indoor environment is the security of the outdoor air intakes. Outdoor air enters the building through these intakes and is distributed throughout the building by the HVAC system. Introducing chemical, biological and radiological agents into the outdoor air intakes allows a terrorist to use the HVAC system as a means of dispersing the agent throughout a

building. Publicly accessible outdoor air intakes located at or below ground level are at most risk—partly due to their accessibility (which also makes visual or audible identification easier) and partly because most chemical, biological and radiological agent releases near a building will be close to the ground and may remain there. Securing the outdoor air intakes is a critical line of defense in limiting an external chemical, biological or radiological attack on a building. See Figure 2-1.

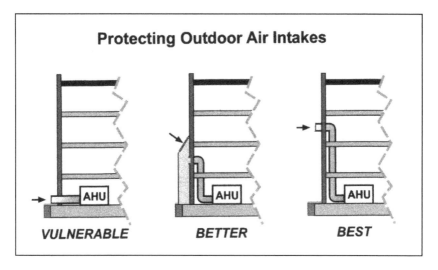

Figure 2-1. Protecting Outdoor Air Intakes
Source: NIOSH

a. *Relocate outdoor air intake vents.* Relocating accessible air intakes to a publicly inaccessible location is preferable. Ideally, the intake should be located on a secure roof or high sidewall. The lowest edge of the outdoor air intakes should be placed at the highest feasible level above the ground or above any nearby accessible level (i.e., adjacent retaining walls, loading docks, and handrails). These measures are also beneficial in limiting the inadvertent introduction of other types of contaminants, such as landscaping chemicals, etc., into the building.

b. *Extend outdoor air intakes.* If relocation of outdoor air intakes is not feasible, intake extensions can be constructed without creating adverse effects on HVAC performance. Depending upon budget, time or the perceived threat, the intake extensions may be temporary or constructed in a permanent, architecturally compatible design. The goal is to minimize public accessibility. In general, this means the higher the extensions, the better— as long as other design constraints (excessive pressure loss, dynamic and static loads) on the structure are appropriately considered. See Figure 2-2. An extension height of 12 feet (3.7 m) will place the intake out of reach of individuals. The entrance to the intake should also be covered with a sloped metal mesh to reduce the threat of objects being tossed into the intake. See Figure 2-3. A minimum slope of 45° is generally adequate.

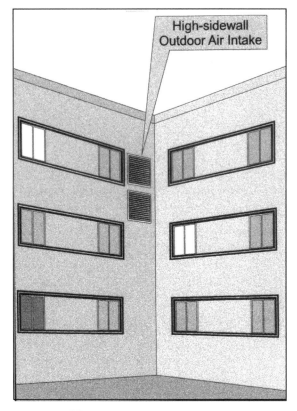

Source: NIOSH

Figure 2-2. High-sidewall Outdoor Air Intake

Extension height should be increased where existing platforms or building features (i.e., loading docks, retaining walls) might provide access to the outdoor air intakes.

c.   *Establish a security zone around outdoor air intakes.* Physically inaccessible outdoor air intakes are the preferred protection strategy. When outdoor air intakes are publicly accessible and relocation or physical extensions are not viable options, perimeter barriers that prevent public access to outdoor air intake areas may be an effective alternative. Iron fencing or similar see-through barriers that will not obscure visual detection of terrorist activities or a deposited chemical, biological

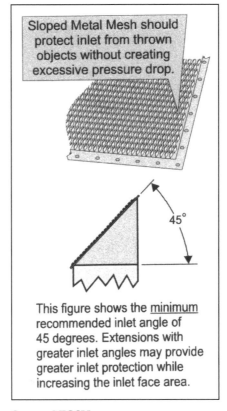

Source: NIOSH

Figure 2-3. Sloped Metal Mesh

or radiological source are preferred. The restricted area should also include an open buffer zone between the public areas and the intake louvers. Thus, individuals attempting to enter these protected areas will be more conspicuous to security personnel and the public. Monitoring the buffer zone by physical security, closed circuit television (CCTV), security lighting, or intrusion detection sensors will enhance this protective approach. See Figure 2-4.

2. **Prevent public access to mechanical areas.** Closely related to the relocation of outdoor air intakes is the security of building mechanical areas. Mechanical areas may exist at one or more locations within a building. These areas provide access to centralized mechanical systems (HVAC, elevator, water, etc.), including filters, air handling units and exhaust systems. Such equipment is susceptible to tampering and may subsequently be used in a chemical, biological

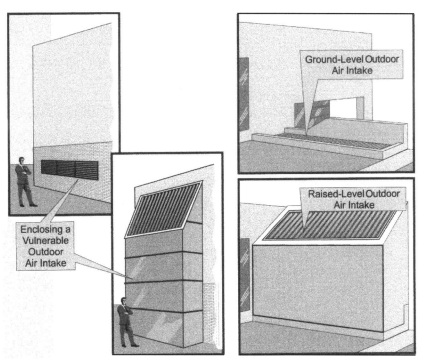

Source: NIOSH

Figure 2-4. Group of 4 outdoor air intakes

and/or radiological attack. Access to mechanical areas should be strictly controlled by keyed locks, keycards or similar security measures. Additional controls for access to keys, keycards and key codes should be strictly maintained.

3. **Prevent public access to building roofs.** Access to a building's roof can allow ingress to the building and access to air intakes and HVAC equipment (e.g., self-contained HVAC units, laboratory or bathroom exhausts) located on the roof. From a physical security perspective, roofs are like other entrances to the building and should be secured appropriately. Roofs with HVAC equipment should be treated like mechanical areas. Fencing or other barriers should restrict access from adjacent roofs. Access to roofs should be strictly controlled through keyed locks, keycards, or similar measures. Fire and life safety egress should be carefully reviewed when restricting roof access.

4. **Implement security measures, such as guards, alarms and cameras to protect vulnerable areas.** Difficult to reach outdoor air intakes and mechanical rooms alone may not stop a sufficiently determined person. Security personnel, barriers that deter loitering, intrusion detection sensors and observation cameras can further increase protection by quickly alerting personnel to security breaches near the outdoor air intakes or other vulnerable locations.

5. **Isolate lobbies, mailrooms, loading docks and storage areas.** Lobbies, mailrooms, including various mail processing areas, loading docks, and other entry and storage areas should be physically isolated from the rest of the building. These are areas where bulk quantities of chemical, biological or radiological agents are likely to enter a building. Building doors, including vestibule and loading dock doors, should remain closed when not in use.

    To prevent widespread dispersion of a contaminant released within lobbies, mailrooms and loading docks, a building's HVAC system(s) should be isolated. Their areas should be maintained at a negative pressure relative to the rest of the building, but at positive pressure relative to the outdoors. Physical isolation of these areas (well-sealed floor-to-roof deck walls, sealed wall penetrations) is critical to maintaining the pressure differential. Physical isolation of these areas also requires special attention to ensure airtight boundaries between these areas and adjacent spaces. In some building designs, establishing a negative pressure differential will present a challenge. A building with lobbies having elevator access is such an example. A qualified HVAC professional can assist in determining if the recommended isolation is feasible for a given building. In addition, lobbies, mailrooms and loading docks should not share a return-air system or return pathway—e.g., ceiling plenum—with other areas of the building. Some of these measures are more feasible for new construction or buildings undergoing major renovation.

Building access from lobby areas should be limited by security checks of individuals and packages prior to their entry into secure areas. Lobby isolation is particularly critical in buildings where the main lobbies are open to the public. Similar checks of incoming mail should also occur before its conveyance into the secure building areas. Side entry doors that circumvent established security checkpoints should be strictly controlled.

6. **Secure return air grilles**. Similar to the outdoor-air intake, HVAC return-air grilles that are publicly accessible and not easily observed by security may be vulnerable to targeting for chemical, biological and/or radiological contaminants. Public access facilities may be the most vulnerable to these types of attacks. A building security assessment can help determine which, if any, protective measures are needed to secure return-air grilles. Caution should be used in selecting various measures so that the performance of the building's HVAC system will not be adversely affected and/or compromised. Some return-air grille protective measures include:
   a. Relocating return-air grilles to inaccessible, yet observable locations;
   b. Increasing security presence—humans or CCTV—near vulnerable return-air grilles;
   c. Directing public access away from return-air grilles; and
   d. Removing furniture and visual obstructions from areas near return air-grilles.

7. **Restrict access to building operation systems by outside personnel.** To deter tampering by outside maintenance personnel,

a building staff member should escort these individuals throughout their service visit and should visually inspect their work before final acceptance of the service. Alternatively, facility managers and building owners can ensure the reliability of pre-screened service personnel from a trusted contractor.

8. **Restrict access to building information**. Information on building operations—including mechanical, electrical, vertical transport, fire and life safety, security system plans and schematics, and emergency operations procedures—should be strictly controlled. Such information should be released to authorized personnel only, preferably by the development of an access list and controlled copy numbering.

9. **General building physical security upgrades**. Physical security measures can be enhanced. For example, security fencing and controlled access points can serve as an initial deterrent. In those buildings that are openly accessible to the public, layered levels of security should be considered. For example, public access to hospital laboratories and ancillary service areas should be restricted to authorized personnel only. Because physical security is of prime concern in open lobby areas, consideration must be given to those facilities with such areas.

VENTILATION AND FILTRATION

HVAC systems and their components should be reevaluated as to how vulnerable the systems and system components are to the introduction of chemical, biological and radiological agents. Relevant issues include the HVAC system controls, the ability of the HVAC system to purge the building, the efficiency of installed filters, the capacity of the system relative to potential filter upgrades, and the significance of uncontrolled leakage into the building. Another consideration is the vulnerability of the HVAC system and components themselves, particularly when the facility is open to the public. For buildings under secure access, interior components may be considered less vulnerable, depending upon the perceived threat and the security measures taken.

1. **Evaluate HVAC control options**. Many central HVAC systems have energy management and control systems that can regulate airflow and pressures within a building on an emergency response basis. Some modern fire alarm systems may also provide useful capabilities during chemical, biological and/or radiological events. In some cases, the best response option, given sufficient warning, might be to shut off the building's HVAC and exhaust systems, thus avoiding the introduction of a biological, chemical or radiological agent from the outside. In other cases, interior pressure and airflow control may prevent the spread of such an agent released in the building. Interior pressure and airflow control can also ensure the safety of egress pathways. NIOSH also recommends that the decision to install emergency HVAC control options should be made in consultation with a qualified HVAC professional. This person must understand the ramifications of various HVAC operating modes on building operation and safety systems.

2. **Assess filtration**. Increasing filter efficiency is one of the few measures that can be implemented in advance to reduce the consequences of both an interior and exterior release of a particulate chemical, biological or radiological agent. However, NIOSH warns, the decision to increase efficiency should be made cautiously, with a careful understanding of the protective limitations resulting from the upgrade. The filtration needs of a building should be assessed with a view to implementing the highest filtration efficiency that is compatible with the installed HVAC system and its required operating parameters. In general, increased filter efficiency will provide benefits to the indoor environmental quality of the building. However, the increased protection from chemical, biological, or radiological aerosols will occur only if the filtration efficiency increase applies to the particle size range and the physical state of the contaminant in question. It is important to note that particulate air filters are used for biological and radiological particles and are not effective for gases and vapors typical of chemical attacks. These types of compounds require adsorbent filters, i.e., activated carbon or other sorbent-type media. Note though, that these filters result in substantial initial and recurring costs.

    As NIOSH notes, upgrading filtration is not as simple as merely replacing a low-efficiency filter with a higher efficiency filter.

Typically, higher efficiency filters have a higher pressure loss, which will result in some airflow reduction through the system. The magnitude of the reduction is dependent on the design and capacity of the HVAC system. If the airflow reduction is substantial, it may result in inadequate ventilation, reductions in heating and cooling capacity, or potentially frozen coils. To minimize pressure loss, deep pleated filters or filter banks having a larger nominal inlet area might be feasible alternatives, if space allows. Also, high-pressure losses can sometimes be avoided by using prefilters or more frequent filter changeouts. Pressure loss associated with adsorbent filters can be even greater.

The integrity of the HVAC system's filter rack or frame system has a major impact upon the installed filtration efficiency. Reducing the leakage of unfiltered air around filters, caused by a poor seal between the filter and the frame, may be as important as increasing filter efficiency. If filter bypass proves to be significant, corrective actions will be needed. Some high-efficiency filter systems have better seals and frames constructed to reduce bypass. It is also recommended that during an upgrade to higher efficiency filters, the HVAC and filtration systems should be evaluated by a qualified HVAC professional to verify proper performance.

While higher filtration efficiency is encouraged and should provide indoor air quality benefits beyond an increased protection from chemical, biological and/or radiological events, the overall cost of filtration should be evaluated. Filtration costs include the periodic cost of the filter media, the labor cost to remove and replace filters, and the fan energy cost required to overcome the pressure loss of the filters. While higher efficiency filters tend to have a higher life cycle cost than lower efficiency filters, this is not always the case. With some higher efficiency filter systems, higher acquisition and energy costs can be offset by longer filter life and a reduced labor cost for filter replacement. Also, improved filtration generally keeps heating and cooling coils cleaner and, thus, may reduce energy costs through improvements in heat transfer efficiency. However, when high efficiency particulate air (HEPA) filters and/or activated carbon adsorbers are used, the overall costs will generally increase substantially.

3.   **Ducted and non-ducted return air systems**. Ducted returns offer limited access points to introduce a chemical, biological or radiological agent. The return vents can be placed in conspicuous

locations, reducing the risk of an agent being secretly introduced into the return system. Non-ducted return air systems commonly use hallways or spaces above dropped ceilings as a return-air path or plenum. Chemical, biological and/or radiological agents introduced at any location above the dropped ceiling in a ceiling plenum return system, will most likely migrate back to the HVAC unit. And, without highly efficient filtration for the particular agent, the agents will be redistributed to occupied areas. Buildings should be designed to minimize mixing between air-handling zones, which can be partially accomplished by limiting shared returns. Where ducted returns are not feasible or warranted, hold-down clips may be used for accessible areas of dropped ceilings that serve as the return plenum. This issue is closely related to the isolation of lobbies and mailrooms, as shared returns are a common way for contaminants from these areas to disperse into the rest of the building. These modifications may be more feasible for new building construction or those undergoing major renovation.

4. **Low-leakage, fast-acting dampers.** Rapid response, such as shutting down an HVAC system, may also involve closing various dampers, especially those controlling the flow of outdoor air in the event of an exterior chemical, biological and/or radiological release. When the HVAC system is turned off, the building pressure compared to outdoors may still be negative. As such, outdoor air is drawn into the building via various leakage pathways, including the HVAC system. NIOSH recommends that consideration should be given to installing low leakage dampers to minimize this flow pathway. Damper leakage ratings are available as part of the manufacturer's specifications. These ratings range from ultra-low to normal categories. Assuming that there is some warning prior to a direct chemical, biological or radiological release, the speed with which these dampers respond to a "close" instruction can also be important. From a protective standpoint, dampers that respond quickly are preferred over dampers that might take 30 seconds or more to respond.

5. **Building air tightness**. Significant quantities of air can enter a building by means of infiltration through unintentional leakage paths in the building envelope. Such leakage is of more concern

for an exterior chemical, biological or radiological release at some distance from a building. For example, this would be true in the event of a large-scale attack as opposed to a sight-specific terrorist act. The reduction of air leakage is a matter of tight building construction in combination with building pressurization. While building pressurization may be a valuable protective strategy in the event of a chemical, biological or radiological attack, it is likely to be more effective in a tight building. However, to be effective, filtration of building supply air must be appropriate for the agent introduced. Although increasing the air tightness of an existing building can be more challenging than during new construction, it should still be seriously considered.

MAINTENANCE, ADMINISTRATION AND TRAINING

Maintenance of ventilation systems and training of staff are critical for controlling exposure to airborne contaminants, such as chemical, biological or radiological agents. In order to ensure proper system maintenance and staff training the following factors must be included:

1. **Emergency plans, policies and procedures**. All buildings should have current emergency plans to address fire, weather and other types of emergencies. In light of past U.S. experiences with anthrax and similar threats, these plans should be updated to consider chemical, biological or radiological attack scenarios and the associated procedures for communicating instructions to building occupants. When developing emergency plans and procedures, it should be noted that there are fundamental differences between chemical, biological and radiological agents. In general, will show a rapid onset of symptoms, while the response to biological and radiological agents will be delayed. Therefore, these plans should identify and address:
   a. Suitable shelter-in-place areas, if they exist
   b. Appropriate use and selection of personal protective equipment, i.e., clothing, gloves, respirators
   c. Directing emergency evacuations
   d. Designated areas and procedures for chemical storage
   e.   control or shutdown, and

f.   Communication with building occupants and emergency responders.

2.   **HVAC maintenance staff training.** Periodic training of HVAC maintenance staff in system operation and maintenance should be conducted. This training would include the procedures to be followed in the event of a suspected chemical, biological and radiological agent release. Training should also cover health and safety aspects for maintenance personnel, as well as the potential health consequences to occupants of poorly performing systems. Development of current, accurate HVAC diagrams and HVAC system labeling protocols should be addressed. These documents can be of great value in the event of a chemical, biological and radiological release.

3.   **Preventive maintenance and procedures.** Procedures and preventive maintenance schedules should be implemented for cleaning and maintaining ventilation system components. Replacement filters, parts and other necessary items should be obtained from known manufacturers and examined prior to installation. It is important that ventilation systems be maintained and cleaned according to the manufacturer's specification. To do this requires information on HVAC system performance, flow rates, damper modulation and closure, sensor calibration, filter pressure loss, filter leakage, and filter change-out recommendations. These steps are critical to ensure that protection and mitigation systems, such as particulate filtration, operate as intended.

INTERNAL AND EXTERNAL RISK FACTORS

Assessing the probability of a terrorist incident, as well as its potential impact, involves identifying the various internal and external risk factors that could precipitate an incident. Identification begins with conducting a comprehensive work-site analysis.

Internal Risk Factors. In its *Handbook for Small Business*, the Occupational Safety and Health Administration (OSHA) provides a detailed checklist that companies of any size may use to identify potential vulnerability. While not all-inclusive, this checklist covers a broad range of internal risk factors, as well as pinpoints key considerations for assessing external risks.

External Risk Factors. Assessing the external risk factors that may indicate vulnerability, presents another challenge to facility managers. These external risk factors are, oftentimes, less obvious and may appear to have little relevance. However, these less obvious threats can pose a significant risk to a company or business entity. For example, a company located near a highway that is used by chemical transporters could be at greater risk than a facility location that is directly adjacent to the plant that manufactures chemicals. The odds for a deliberate vehicular attack/accident involving a chemical transporter increase the risk potential for that first company.

Other external risk factors that must be included in any risk assessment are:

- The actual physical location of the facility or building (urban versus suburban geographic location, including the concomitant population density of same);

- Proximity to protective and emergency medical services;

- Climate/weather features that are unique to the geographic location of the facility or building.

Each of these factors, in turn, should be evaluated on the basis of whether any one, or a combination of factors, would either increase or decrease the likelihood of a chemical, biological or radiological incident. These factors should also be evaluated on the basis of whether or not there would be a negligible effect or have a negligible effect on the likelihood of such an incident occurring.

Sources

Gustin, Joseph F. *Disaster and Recovery Planning: A Guide for Facility Managers, 3rd ed.*, Lilburn, GA: The Fairmont Press, Inc., 2004.

Gustin, Joseph F. *The Facility Manager's Handbook*, Lilburn, GA: The Fairmont Press, Inc., 2003.

Gustin, Joseph F. *Safety Management: A Guide for Facility Managers*, New York: UpWord Publishing, Inc., 1996.

National Institute of Occupational Safety and Health, NIOSH. *Protection from Chemical, Biological, or Radiological Attacks*, DHHS (NIOSH) Publication No. 2002-139. Cincinnati, Ohio, 2002.

Chapter 3

The Role of HVAC in a Chemical, Biological or Radiological Attack

Targets for terrorists wishing to incite panic or disrupt either a local or national economy, etc., include any public, private or government building. Targets also could include offices, laboratories, schools, bus/airline/train terminals, hospitals, retail facilities and public gatherings such as shopping centers/malls, sports arenas, national monuments and historic sites.

In addition to integrating people with their physical environment, facility managers and building owners have a responsibility to protect their employees/occupants from harm. Since there is uncertainty as to which building could be a likely target, it is up to each facility manager and building owner to decide the risk level of their building. Once the risk level has been decided, implementation of protective measures should be based on the following—perceived risk for the building and its employee/occupants, engineering and architectural feasibility and cost. Deciding upon the level of building protection from an airborne chemical, biological or radiological attack is a priority for the facility manager and building owner. Making the building/facility into a less attractive target should be a priority.

While it is uncertain to predict the time and place in which a building can be targeted with an airborne attack, there are certain steps that facility managers and building owners can take to protect their employees/occupants from an airborne chemical, biological or radiological attack. The vulnerability of a building's HVAC system to a chemical, biological or radiological threat is of particular concern to facility managers and building owners. Because the system components can become entry points and/or distribution points for such agents, HVAC systems must be properly designed, installed and maintained. A properly designed, installed and maintained HVAC system can reduce the effects of a chemical, biological or radiological release either outside or within the building. This is accomplished by removing any hazardous contaminants from the building's air supply. As such, facility managers need reliable information about filtration and air-cleaning options. For example, facility managers need to know the:

☐   types of air-filtration and air-cleaning systems that are effective for various chemical, biological or radiological contaminants;

    ☐   types of air-filtration and air-cleaning systems that can be incorporated into an existing HVAC system;

    ☐   types of air-filtration and air-cleaning systems that can be incorporated into existing buildings when they undergo comprehensive renovation; and

    ☐   proper maintenance procedures for maintaining the air-filtration and air-cleaning systems installed in a building.

After this information is obtained, facility managers and building owners can make the necessary decisions regarding appropriate filtration systems, as well as decisions for any other related protective measures.

AIR-FILTRATION / AIR-CLEANING SYSTEMS

Air-filtration and air-cleaning systems can remove contaminants from a building's environment. The effectiveness of a particular filter or "air-cleaner" depends upon the nature of the contaminant—aerosol v. gases. According to the National Institute for Occupational Safety and Health (NIOSH), air filtration refers to the removal of aerosol contaminants from the air, while air cleaning refers to the removal of gases or vapors from the air. Airborne contaminants are gases, vapors or aerosols, which are small solid and liquid particles. It is important to note that sorbents collect gases and vapors, but not aerosols. On the other hand, particulate filters remove aerosols, but not gases and vapors. The ability of a given sorbent to remove a contaminant depends upon the characteristics of the specific gas or vapor, as well as other related factors. The efficiency of a particulate filter to remove aerosols depends upon particle size, in combination with the type of filter used and HVAC operating conditions. Larger-sized aerosols can be collected on lower-efficiency filters, but the effective removal of a small-sized aerosol requires a higher-efficiency filter.

In addition to proper filter or sorbent selection, several other issues must be considered before filtration systems are installed or upgraded. These issues include filter bypass, cost and the building envelope.

- *Filter bypass* is a common problem found in many HVAC filtration systems. Filter bypass occurs when air moves around the filter, rather than through the filter. This decreases collection efficiency and defeats the intended purpose of the filtration system. Filter bypass is often caused by poorly fitting filters, poor sealing of filters in their framing systems, missing filter panels, or leaks and openings in the air-handling unit between the filter bank and blower. By simply improving filter efficiency without addressing filter bypass little, if any, benefit is provided.

- *Cost* is another issue affected by HVAC filtration systems. Life-cycle cost should be considered, i.e., initial installation, replacement, operation, maintenance, etc. Not only are higher-efficiency filters and sorbent filters more expensive than the commonly used HVAC system filters, but also fan units may need to be changed to handle the increased pressure drop associated with the upgraded filtration systems. Although improved filtration will normally come at a higher cost, costs can be partially offset by the accrued benefits. These benefits include cleaner and more efficient HVAC components and improved indoor environmental/air quality.

- The *building's envelope* matters. Filtration and air cleaning affect only the air that passes through the filtration and air-cleaning device, whether it is outdoor air, re-circulated air, or a mixture of the two. Outside building walls in residential, commercial and institutional buildings are quite leaky, and the effect from negative indoor air pressures (relative to the outdoors) allows significant quantities of unfiltered air to infiltrate the building envelope. Field studies have shown that, unless specific measures are taken to reduce infiltration, as much air may enter a building through infiltration (unfiltered) as through the mechanical ventilation (filtered system). Therefore, filtration alone will not protect a building from an outdoor chemical, biological and/or radiological release. This is particularly so for systems in which no make-up air or inadequate overpressure is present. Instead, air filtration must be considered in combination with other steps, such as building pressurization and envelope air tightness. By doing so, the likelihood that the air entering the building actually passes through the filtration and air cleaning system is increased.

Chemical, biological and/or radiological agents may travel in the air as a gas or an aerosol. Chemical warfare agents with relatively high vapor pressure are gaseous, while many other chemical warfare agents could potentially exist in either state. Biological and radiological agents are largely aerosols.

Health Consequences

Some health consequences from chemical, biological and/or radiological agents are immediate, while others may take much longer to appear. Chemical, biological and/or radiological agents, e.g., arsine, nitrogen mustard gas, anthrax, and radiation from a dirty bomb, can enter the body through a number of routes including inhalation, skin absorption, contact with eyes or mucous membranes, and ingestion. The amount of a chemical, biological and/or radiological agent required to cause specific symptoms varies among agents; however, these agents are generally much more toxic than common indoor air pollutants. In many cases, exposure to extremely small quantities may be lethal. Symptoms are markedly different for the different classes of agents (chemical, biological or radiological). Symptoms resulting from exposure to chemical agents tend to occur quickly. Most chemical warfare agents—gases—are classified by their physiological effects, e.g., nerve, blood, blister, and choking. Toxic industrial chemicals (TICs) can also elicit similar types of effects. Conversely, symptoms associated with exposure to biological agents—bacteria, viruses—vary greatly with the agent and may take days or weeks to develop. These agents may result in high morbidity and mortality rates among the targeted population. Symptoms from exposure to ionizing radiation can include both long-term and short-term effects.

FILTRATION AND AIR-CLEANING PRINCIPLES

Simply stated, filtration and air cleaning remove unwanted material from an air stream. For HVAC applications, this involves air filtration and, in some cases, air cleaning—for gas and vapor removal. The collection mechanisms for particulate filtration and air cleaning systems are very different. The following description of the principles governing filtration and air cleaning briefly provides an understanding of the most important factors facility managers should consider when selecting or enhancing the building's filtration system.

Particulate Air Filtration

Particulate air filters are classified as either mechanical filters or electrostatic filters, i.e., electrostatically enhanced filters. Although there are many important performance differences between the two types of filters, both are fibrous media and used extensively in HVAC systems to remove particles, including biological materials, from the air. A fibrous filter is an assembly of fibers that are randomly laid perpendicular to the airflow.

Fibrous filters of different designs are used for various applications. *Flat-panel filters* contain all of the media in the same plane. This design keeps the filter face velocity and the media velocity roughly the same. When *pleated filters* are used, additional filter media are added to reduce the air velocity through the filter media. This enables the filter to increase collection efficiency for a given pressure drop. Pleated filters can run the range of efficiencies from a minimum efficiency reporting value (MERV) of 6 up to and including high-efficiency particulate air (HEPA) filters. With *pocket filters*, air flows through small pockets or bags constructed of the filter media. These filters can consist of a single bag or have multiple pockets, and an increased number of pockets increases the filter media surface area. As in pleated filters, the increased surface area of the pocket filter reduces the velocity of the airflow through the filter media, allowing increased collection efficiency for a given pressure drop. *Renewable filters* are typically low-efficiency media that are held on rollers. As the filter loads, the media are advanced or indexed, providing the HVAC system with a new filter. There are four different collection mechanisms that govern particulate air filter performance: inertial impaction, interception, diffusion and electrostatic attraction. The first three of these mechanisms apply mainly to mechanical filters and are influenced by particle size.

- Impaction occurs when a particle traveling in the air stream and passing around a fiber, deviates from the air stream—due to particle inertia—and collides with a fiber.

- Interception occurs with a large particle and, because of its size, collides with a fiber in the filter that the air stream is passing through.

- Diffusion occurs when the random motion of a particle causes that particle to contact a fiber.

- Electrostatic attraction, the fourth mechanism, plays a very minor role in mechanical filtration. After fiber contact is made, smaller particles are retained on the fibers by a weak electrostatic force.

- Electrostatic filters contain electrostatically enhanced fibers, which actually attract the particles to the fibers, in addition to retaining them. Electrostatic filters rely on charged fibers to dramatically increase collection efficiency for a given pressure drop across the filter.

> Electrostatically enhanced filters are different from electrostatic precipitators, also known as electron air cleaners. Electrostatic precipitators require power and charged plates to attract and capture particles.

As mechanical filters load with particles over time, their collection efficiency and pressure drop typically increase. Eventually, the increased pressure drop significantly inhibits airflow and the filters must be replaced. For this reason, pressure drop across mechanical filters is often monitored because it indicates when to replace filters.

Conversely, electrostatic filters, which are composed of polarized fibers, may lose their collection efficiency over time or when exposed to certain chemicals, aerosols or high relative humidity. Pressure drop in an electrostatic filter generally increases at a slower rate than it does in a mechanical filter of similar efficiency. Thus, unlike the mechanical filter, pressure drop for the electrostatic filter is a poor indicator of the need to change filters. The differences between mechanical and electrostatic filters should be noted when selecting the appropriate HVAC filter because of the impact they will have on the filters performance, i.e., collection efficiency over time, as well as on maintenance requirements which determine change-out schedules.

Gas-Phase Air Cleaning

Some HVAC systems may be equipped with sorbent filters, designed to remove pollutant gases and vapors from the building environment. Sorbents use one of two mechanisms for capturing and controlling gas-phase air contaminants: physical adsorption and chemisorption. Both capture mechanisms remove specific types of gas-phase contaminants from indoor air. Unlike particulate filters, sorbents cover a wide range of highly porous materials varying from simple clays and carbons to complex engineered polymers. Many sorbents—not including those that are chemically active—can be regenerated by application of heat or other processes.

Understanding the precise removal mechanism for gases and vapors is often difficult due to the nature of the adsorbent and the processes involved. While knowledge of adsorption equilibrium helps in understanding vapor protection, sorbent performance depends on such properties as mass transfer, chemical reaction rates and chemical reaction capacity.

RECOMMENDATIONS:
FILTER AND SORBENT SELECTION,
OPERATIONS, UPGRADE AND MAINTENANCE

Before selecting a filtration and air-cleaning strategy that includes a potential upgrade in response to perceived types of threats, an understanding of the building and its HVAC systems must be developed. A vital part of this effort will be to evaluate the total HVAC system. The design of the HVAC system must be assessed in terms of its intended operation and then compared to how it actually operates. In large buildings, this evaluation is likely to involve many different air-handling units and system components.

Initially, several questions will need to be answered. Many of these questions may be difficult to answer without the assistance of qualified professionals, i.e., security specialists, HVAC engineers, industrial hygienists, etc. This assistance can help determine potential threat to ventilation/filtration as well as to indoor air quality. The answer to these questions should serve as a guide in the decision making process regarding the following factors:

•   types of filters and/or sorbents that should be installed in the HVAC system;

- efficiency of filters and/or sorbents; and

- the procedures to be developed for system maintenance.

Because of the wide range of building and HVAC system design, no single, off-the-shelf system can be installed in all buildings to protect against all chemical, biological and/or radiological releases. While some system components could possibly be used in a number of buildings; the systems should be designed specifically for each building, as well as for each application. Some of the important questions that must be asked include:

- How are the filters in each system held in place and how are they sealed? Are the filters simply held in place by the negative pressure generated from downstream fans? Do the filter frames—the part of the filter that holds the filter media—provide for an air-tight, leak-proof seal with the filter rack system, that part of the HVAC system that holds the filters in place?

- What types of air contaminants are of concern? Are the air contaminants particulate, gaseous, or both? Are they TICs, toxic industrial materials (TIMs), or military agents? How toxic are they? Check with the local emergency or disaster planning body to determine if there are large quantities of TICs or TIMs near the building location or if there are specific concerns about military, chemical or biologic agents.

- How can the chemical, biological and/or radiological agents enter the building? Are they likely to be released internally or externally to the building envelope, and how can various release scenarios be addressed?

- What is needed? Are filters or sorbents needed to improve current indoor air quality, provide protection in an accidental or intentional release of a nearby chemical processing plant, or provide protection from a potential terrorist attack using chemical, biological and/or radiological agents?

- How clean does the air need to be for the occupants, and how much can be spent to achieve that desired level of air cleanliness? What are the total costs and benefits associated with the various levels of filtration?

• What are the current system capacities (i.e., fans, space for filters, etc.), and what is needed? What are the minimum airflow needs for the building?

• Who will maintain these systems and what are the capabilities of this person(s)?

It is important to recognize that improving building protection is not an all or nothing proposition. Because many chemical, biological and/or radiological agents are extremely toxic, high contaminant removal efficiencies are needed. However, many complex factors can influence the human impact of a chemical, biological and/or radiological release—i.e., agent toxicity, physical and chemical properties, concentration, wind conditions, means of delivery and release location. Incremental improvements to the removal efficiency of a filtration or air-cleaning system are likely to lessen the impact of a chemical, biological and/or radiological attack to a building environment, as well as to the building occupants. Additionally, indoor air quality is enhanced with such improvements.

A Word about Filter or Sorbent Bypass and Air Filtration
Ideally, all airflow should pass through the installed filters of the HVAC system. However, filter bypass is a common problem. Filter bypass occurs when air flows around a filter or through some other unintended path. Preventing filter bypass becomes more important as filter collection efficiency and pressure drop increase. Airflow around the filters result from various imperfections, e.g., poorly sealed filters, which permit particles to bypass the filters, rather than passing directly into the filter media. Filters can be held in place with a clamping mechanism, but this method may not provide an airtight seal. The best high-efficiency filtration systems have gaskets and clamps that provide an airtight seal. Any deteriorating or distorted gaskets should be replaced and checked for leaks. Visual inspection of filters for major leakage around the edges can be done. The best method of checking for leaks involves a particle counter or aerosol photometer. Merely placing a light source behind the filter is not adequate. Finally, no faults or other imperfections should exist

within the filter media, and performance should be evaluated using a quantitative test or measurement.

Another issue to consider is infiltration of outdoor air into the building. Air infiltration may occur through openings in the building envelope—such as doors, windows, ventilation openings and cracks. Typical office buildings are quite porous. To achieve the most effective filtration and air cleaning system against external chemical, biological and / or radiological threats, outdoor air leakage into the building must be minimized. Dramatically reducing leakage can be impractical for many older buildings, which may have large leakage areas, operable windows, and decentralized HVAC systems. In these instances, other protective measures, such as those outlined in the NIOSH *Guidance for Protecting Building Environments from Airborne Chemical, Biological, or Radiological Attacks*, should be considered.

Initially, the decision must be made as to which portions of the building should be included in the protective envelope. Areas requiring high air exchange, such as some mechanical rooms, may be excluded. To maximize building protection, reduce the infiltration of unfiltered outdoor air by increasing the air tightness of the building envelope— eliminating cracks and pores—and introducing enough filtered air to place the building under positive pressure with respect to the outdoors. It is much easier and more cost efficient to maintain positive pressure in a building if the envelope is tight, so it is recommended to use these measures in combination. The U.S. Army Corps of Engineers recommends that for external terrorist threats, buildings should be designed to provide positive pressure at wind speeds up to 12 km / hr (7 mph). Designing for higher wind speeds will give even greater building protection [U.S. Army 1999].

In buildings that have a leaky envelope, maintaining positive indoor pressure may be difficult to impossible. Interior/exterior differential air pressures are in constant flux due to wind speed and direction, barometric pressure, indoor/outdoor temperature differences—stack effect—and building operations, such as elevator movement or HVAC system operation. HVAC system operating mode is also important in maintaining positive indoor pressure. For example, many HVAC systems use an energy savings mode on the weekends and at night to reduce outside air supply and, hence, lower building pressurization. In cold climates, an adequate and properly positioned vapor barrier must be ensured before any pressurization to minimize condensation is attempted. This is important to prevent mold and other problems. All of these factors—leaky envelope, negative indoor air pressure, energy savings mode—influence building air infiltration and must be considered when the building is tightened. Building pressurization or tracer gas testing can be used to evaluate the air tightness of the building envelope.

RECOMMENDATIONS:
OPERATIONS AND MAINTENANCE

Filter performance depends on proper selection, installation, operation, testing and maintenance. The scheduled maintenance program should include procedures for installation, removal and disposal of filter media and sorbents. Only adequately trained personnel should perform filter maintenance and only while the HVAC system is not operating— locked out/tagged out—to prevent contaminants from entering the moving air stream.

Recommendation: Following a chemical, biological and/or radiological release, appropriate emergency response and/or health and safety professionals must be consulted before attempting HVAC system maintenance.

If a chemical, biological and/or radiological release occurs in or near the building, significant hazards may be present within the building's HVAC system. Following a chemical, biological and/or radiological release, contaminants will have collected on the system components, the particulate filters, or within the sorbent bed itself. As such, these accumulated materials present a hazard to personnel servicing the various systems. Therefore, before servicing these systems following a release, consult with the appropriate emergency response and/or health and safety professionals to develop a plan for returning the HVAC systems and building to service. Because of the wide variety of buildings, contaminants and scenarios, it is not possible to provide a generic plan. However, a site specific plan should include requirements for personnel training and appropriate personal protective equipment.

Recommendation:
Understand how filter type affects change-out schedules.

Proper maintenance, including the monitoring of filter efficiency and system integrity, is critical to ensuring HVAC systems operate as intended. The change-out schedule for various filter types may be significantly different. One reason for differences is that little change in pressure drop occurs during the loading of an electrostatic filter, as opposed to mechanical filters. Ideally, optical particle counters or other quantitative measures of collection efficiency should be used to determine the change-out schedule for electrostatic filters. Collecting objective data—experimental measurements—will allow for an optimization of electrostatic filter life and filtration performance. The data should be particle-size selective. In that way filtration efficiencies that are based on particle size—e.g., micrometer, sub-micrometer and most penetrating size—can be determined. On the other hand, mechanical filters show larger pressure drop increases during loading, and hence, pressure drop can be used to determine appropriate change-out schedules. If using mechanical filters, a manometer or other pressure-sensing device should be installed in the mechanical filtration system to provide an accurate and objective means of determining the need for filter replacement. Pressure drop characteristics of both mechanical and electrostatic filters are supplied by the filter manufacturer.

Recommendation:
Ensure that maintenance personnel are well trained.

Qualified individuals should be responsible for the operation of the HVAC system. These individuals must have a general working knowledge of the HVAC system and its function since they are responsible for monitoring and maintaining the system. Because system monitoring and maintenance includes filter change-out schedules, documentation and record-keeping, the HVAC person should also be involved in the selection of the appropriate filter media for a given application. It is important to note that due to the sensitive nature of HVAC systems, appropriate background checks should be completed and assessed for any personnel who have access to the system and equipment.

Recommendation:
Handle filters with care and inspect for damage.

Mechanical filters are often made of glass fibers. As such, they are relatively delicate and should be handled carefully to avoid damage. Filters enclosed in metal frames are heavy and may cause problems because their additional weight may place stress on the filter racks. This increased weight may require a new filter support system that has vertical stiffeners and better sealing properties to ensure total system integrity. Polymeric electrostatic filters are more durable and less prone to damage than mechanical filters.

To prevent installation of a filter that has been damaged in storage or one that has a manufacturing defect, all filters should be checked before installation. The seams of these filters should also be visually inspected for total integrity. The filters should be held in front of a light source and checked for voids, tears or gaps in its media and frames.

Recommendation:
Personal Protective Equipment

Authorized personnel performing maintenance and filter replacement on any ventilation system that is likely to be contaminated should use appropriate personal protective equipment. Guidelines for the personal protective equipment (respirators, gloves, etc.) have been established by the Occupational Safety and Health Administration (OSHA). See standard 29 Code of Federal Regulations (CFR)1910.132 and 1910.134. Chapter 7 outlines regulatory requirements.

Maintenance and filter change-out should be performed only when a system is shut down to avoid re-entrainment and system exposure. Old filters should be placed in sealed bags upon removal and where feasible, particulate filters may be disinfected in a 10 percent bleach solution or other appropriate biocide before removal. It is important to note that the HVAC system must be shut down when disinfecting compounds are used. An equally important consideration is to ensure that the disinfecting compounds used are compatible with the HVAC system components. Because decontaminating filters exposed to chemical, biological and/or radiological agents require knowledge of the type of agent, local hazmat teams and contractors, themselves, should be able to provide safety-related information concerning the decontaminating compounds. Additionally they should provide detailed information on the proper hazardous waste procedures.

SUMMARY

Filtration and air-cleaning systems may protect a building and its occupants from the effects of a chemical, biological and/or radiological attack. Although it is impossible to completely eliminate the risk from an attack, filtration and air-cleaning systems are important components of a comprehensive plan to reduce the consequences. Chemical, biological and/or radiological agents can effectively be removed by a properly designed, installed and well-maintained filtration and air-cleaning system. These systems have other benefits besides reducing clean-up costs and delays, should an event occur. These benefits include: improving building cleanliness, improving HVAC system efficiency, potentially preventing cases of respiratory infection, reducing exacerbations of asthma and allergies, and generally improving building indoor air quality. Poor indoor air quality has also been associated with eye, nose and throat irritation, headaches, dizziness, difficulty concentrating and fatigue.

Initially, the facility manager/building owner must fully understand the design and operation of the building and its HVAC system. Backed with that knowledge, along with an assessment of the current threat and the level of protection needed from the system, informed decisions can be made regarding the building's filtration and air-cleaning needs. In some situations, the existing system may be adequate. In other situations, major changes or improvements may be warranted.

In most buildings, mechanical filtration systems for aerosol removal are more common than sorbents for gas and vapor removal. Decisions regarding collection efficiency levels of particulate filters should be made in accordance with government/agency standards (e.g., ASHRAE standards 52.1 and 52.2). Selection of the most appropriate sorbents for gaseous contaminants is more complex. To optimize effectiveness, air infiltration should be minimized and filter bypass should be eliminated. Maintenance plans and operations should ensure that the system functions as intended for long periods. Life-cycle analysis ensures that filtration and air-cleaning options satisfies building needs as well as providing protection to building occupants.

Sources

National Institute for Occupational Safety and Health, *Guidance for Filtration and Air-Cleaning Systems to Protect Building Environments from Airborne Chemical, Biological, or Radiological Attacks*, DHHS (NIOSH) Publication No. 2003-136. April 2003.

Chapter 4

Safeguarding Buildings

W hile physical security is important to the safety of the building's occupants, focus must also be directed towards the protection of the building environment from airborne chemical, biological or radiological attacks. This focus is accomplished through the use of filtration and air-cleaning systems. As such, the emphasis is placed on preventing an attack rather than recovering from an attack.

Building owners, facility managers and others including architects and designers, should have reliable information about filtration and air-cleaning options. The National Institute of Occupational Safety and Health (NIOSH) lists the areas that building owners and facility managers need to know:

- The types of air-filtration and air cleaning types that are effective for various chemical, biological and radiological agents;

- The types of air-filtration and air-cleaning systems that can be implemented in an existing HVAC system;

- The types of air-filtration and air-cleaning systems that can be incorporated into existing buildings when they undergo comprehensive renovation; and

- Proper air filtration and air-cleaning maintenance procedures, combined with other various protective measures.

For the purposes of this discussion, air filtration refers to the removal of aerosol contaminants from the air, while air cleaning refers to the removal of gases or vapors from the air. Airborne contaminants are gases, vapors, or aerosols (small, solid and liquid particles). It is important to realize that sorbents collect gases and vapors, but not aerosols. On the other hand, particulate filters remove aerosols, but not gases and vapors.

FACTS ABOUT AIRBORNE HAZARDS

Most hazardous chemicals have warning properties. For example, their vapors or gases can be perceived by the senses. According to the U.S. Army Corps of Engineers, soldiers in World War I and World War II were taught to identify by smell such agents as mustard, phosgene and chlorine. This detection method—the sense of smell—proved effective for determining when to put on and take off a gas mask. In addition to the sense of smell, hazardous chemicals can also be detected by sight, taste, eye irritation, skin irritation or respiratory tract irritation. These detection "methods" are useful indicators to the existence of vapors or gases before serious physical effects can occur. However, it is important to note that there are some toxic industrial chemicals which are imperceptible—i.e., they cannot be detected or perceived by any of the senses. The most common of these chemicals is carbon monoxide. Because it is odorless and colorless, carbon monoxide causes many deaths in buildings each year.

Biological agents, like carbon monoxide, are also imperceptible. They cannot be detected by human senses. There are no detection devices that can determine the presence of biological agents in the air. However, airborne hazards can be detected by observation. By observing symptoms or effects in others, one can be alerted to a biological hazard. Some of the warning signs include dizziness, nausea and dead animals. Other warning signs may involve seeing and hearing something out of the ordinary, such as the hissing sound that is caused by a rapid release from a pressurized cylinder. Awareness to warning properties, including the signs and symptoms in other people, is the basis of a protective action plan. Such a plan applies for possible protective actions: sheltering in place, using protective masks, evacuating and purging.

When planning for imperceptible airborne hazards such as the presence of carbon monoxide and biological agents, the protective measures are most likely in place at a building or facility. Those protective measures include: filtering all air brought into the building on a continuous basis; and using automatic real-time sensors that are capable of detecting the imperceptible agents.

Types of Releases

There are two general types of releases—external and internal. External releases can result from storage or transport accidents, fires or malicious acts. With an outdoor (external) release, the source of the hazard is most likely to be at or near ground level. When gases or aerosols are released at ground level, they tend to remain at ground level under stable conditions. These stable conditions normally occur at night, dusk, dawn and on overcast days. On sunny days, when the ground is hotter than the air above it, plumes tend to spread upward and be diluted as they rise. The plumes originating at ground level will be diverted upward as they travel over buildings. In general, a plume will take the shortest path past a building. If the width of a building is more than twice its height, the shortest path will be over the building and the plume will travel upward to openings on upper floors.

The potential for an internal release of hazardous materials is determined by:

1) the presence of hazardous materials stored in the building;
2 security measures to prevent hazardous materials from being brought into the building; and
3) architectural and mechanical features to isolate or limit the spread of hazardous material if an internal release occurs.

Hazardous materials can be carried into a building by people or in the delivery of mail, supplies and equipment. Therefore, the likelihood of an internal release is also determined by the accessibility of the building to the public and the presence of entry screening measures for people, mail and supplies.

SAFEGUARDING BUILDINGS: IMMEDIATE ACTIONS

According to the Lawrence Berkeley National Laboratory, there are several actions that can be taken to safeguard a building. These include:

• Identifying fresh air intakes and preventing unauthorized access;

• Securing mechanical (HVAC) room doors to prevent unauthorized access;

• Securing building plans and HVAC plans from unauthorized access;

- Developing an emergency response team and establishing operational procedures; and

- Planning and practicing separate emergency response procedures for indoor and outdoor releases of chemical/biological agents.

Identifying Fresh Air Intakes and
Preventing Unauthorized Access

The building's ventilation system can become quickly contaminated with a chemical or biological agent. A terrorist can gain access to the interior of the building from the outside by utilizing an accessible air intake. An air intake at a higher elevation such as the roof reduces its vulnerability, i.e., terrorist accessibility.

Security for a vulnerable air intake should include, but not be limited to, fencing or restrictions around the air intake. A container filled with anthrax spores can be easily thrown into an air intake from a parking lot or across the street. Baffles over the air intake can help mitigate this action, but should be done only with careful evaluation. Installing baffles could also affect energy efficiency and the amount of outside air that could be pulled in.

It is important to note that many buildings take in large amounts of air through the building exhausts. This phenomenon occurs during warm weather, particularly in buildings that do not have a return fan or buildings that do not use volume-matching to control the return fans. Since building designers assume that there will be no inward flow through the exhausts, exhaust registers are often placed on loading docks or adjacent to sidewalks. As such, they may be particularly vulnerable. Facility managers must determine whether their buildings are subject to this phenomenon. If the registers are placed on loading docks, or if they are adjacent to sidewalks, consideration must be given to restricting access to the exhausts, or at the least, modifying the return fan controls. If there is not significant inward flow through the exhaust registers, biological or chemical agents can be prevented from entering this route.

Securing Mechanical (HVAC)
Room Doors to Prevent Unauthorized Access

Rooms that house HVAC equipment should be locked and keyed. Only those staff members who need access to the HVAC room should be able to gain access to that room. Doing so can prevent a terrorist from gaining access to the building's HVAC equipment and contaminating either the building or ventilation zone with a chemical or biological agent.

Securing Building Plans and HVAC Plans from Unauthorized Access

Contaminating a building, maximizing casualties or targeting specific individuals through an indoor chemical or biological attack is an easy task for a terrorist with access to the building's plans and/or the HVAC plans. Therefore, it is imperative that building plans and ventilation system details should be made available only to authorized people. If either the building plans or the HVAC plans are provided to any authorized contractor for building work, these plans should be recovered after use.

Developing an Emergency Response Team and Establish Operational Details

Since emergencies require rapid response, an emergency action team of employees and a back-up team should be created. This team and its back-up should have well-defined responsibilities. These responsibilities include:

- Decision making about the following critical areas:
  — Building evacuation
  — HVAC system operation
- Contacting authorities
- Providing instructions to building occupants
- Manipulating the HVAC system as needed
- Coordinating first aid

Planning and Practicing Separate Emergency Response Procedures for Indoor and Outdoor Releases of Chemical/Biological Agents

Facility managers and building owners should understand the differences in the best responses to indoor versus outdoor releases, and should practice the steps needed to be taken in each case. The first response to an outdoor chemical/biological release should include shutting down the building's ventilation system and closing all doors and windows. Generally speaking, the response to an indoor chemical/biological release should include evacuation.

SAFEGUARDING BUILDINGS: LONG TERM ACTIONS

The following lists some of the long term actions that can help make a building safer in the event of a chemical or biological release. The actions that should be performed depend, of course, upon the costs and level of threat to, or upon the particular building:

1. Ensure that building operators can quickly manipulate the HVAC control systems to respond to different kinds of attack;

2. Upgrade HVAC particle filters and seal gaps to prevent air bypass;

3. Establish internal "safe zones" for sheltering-in-place, and exterior locations for evacuation; for an outdoor release, people should remain indoors;

4. Prevent the air from mail rooms and other high-risk locations from entering into the rest of the building; and

5. Weatherize the building to reduce outdoor air penetration.

**Ensure that Building Operators can Quickly Manipulate
the HVAC Control Systems to Respond to Different Kinds of Attacks**
It should be possible to immediately shut off the HVAC system—including closing dampers that admit outdoor air and closing exhaust dampers—or to put the system on 100% fresh (outdoor) air. Dampers should be checked for leakage and replaced or repaired if necessary.

To ensure that building operators know which supply and exhaust systems serve which areas, those areas served by different air handling units should be surveyed and clearly shown on floor plans. An updated copy of the ventilation plans in the room/rooms from which the HVAC system can be controlled should be maintained.

Optimally, there should be more than one secure location from which the entire HVAC system can be controlled. At minimum, though, there must be at least one secure location. By doing so, it precludes the necessity for an emergency team member or authorized personnel to move through a contaminated building in order to manipulate the system.

**Upgrade HVAC Particle Filters and
Seal Gaps to Prevent Air Bypass**
Most building HVAC systems have some type of particle filter. Substituting a more effective filter—particularly for small particle sizes—

can reduce the risk of spreading a biological agent through the building via the HVAC system.

Since more effective filters can lead to significant pressure drop, the use of deep pleated filters or filter banks with a larger inlet area can be used. This, of course, is premised upon the space available. The use of these types of filters can ensure that any unfiltered air around the filter or through leaks and ducts that occur between the filter and the fan, can be minimized or eliminated.

Improved filters can provide significant protection from a biological release, but these filters should be installed correctly. Ducts should also be sealed to reduce air bypass.

Careful evaluation should be done before installing new filters for several reasons. These new filters may require changes in other equipment in order to maintain air quality and comfort. And, these new filters may also increase energy use, as well.

For an Outdoor Release, People Should Remain Indoors
Establish internal "safe zones" for sheltering-in-place, and exterior locations for evacuations. "Shelter-in-place" rooms can be created or identified. These rooms provide a shelter in the event of an outdoor release. The goal of a shelter-in-place is to create areas where outdoor air infiltration is very low; usually such rooms will be in the inner part of the building with no outside windows. They should have doors that are fairly effective at preventing airflow from the hallways; i.e., there should be no gap or only a very small gap at the bottom of the door. Sliding doors are more effective for a shelter-in-place since opening and closing a conventional door can pump large amounts of air into the room. Since sliding doors can reduce the large amounts of air being pumped into the room, considerations should be given to replacing conventional doors with sliding doors. Since bathrooms often have an exhaust duct that leads directly to the outside, they are usually a bad choice for a shelter-in-place. If the bathroom exhaust fan is left on, air will be drawn from another part of the building into that bathroom contaminating that room. If the exhaust fan is turned off, then the duct can allow outside air to directly enter the bathroom. Exhaust fans for bathrooms and utility rooms are often controlled separately from the HVAC system.

Also, purified air may be provided to the safe area by making modifications to the HVAC system. These modifications include adding filters and an air supply dedicated to the safe area.

For an Indoor Release, People Should Exit the Building

The chemical/biological agent will be carried out of the building through windows, doors and vents, so people should congregate *upwind* of the building.

At least two different evacuation zones should be identified in advance, and the appropriate zone should be used depending upon the wind direction.

Do Not Mix Air from Mail Rooms and Other
High-risk Locations into the Rest of the Building

The most likely locations for introducing toxic substances into a building include mailrooms, areas with public access and delivery areas such as loading docks. If the HVAC systems for these areas do not mix air into the rest of the building, the spread of the agent will be greatly reduced. Mixing into the general building air can be prevented by providing a separate air-handling unit for these areas or by eliminating return air for these areas and exhausting them directly. Contamination may still spread along hallways, etc., but the contamination will be much slower.

Adjusting the HVAC supply and exhaust so that the high-risk areas are slightly depressurized with respect to the rest of the building may also be considered. In that way, air will flow from other areas into the high-risk areas rather than the other way around.

Weatherize the Building to Reduce Outdoor Air Penetration

Cracks around windows and doors allow conditioned air to escape the building, and outdoor air to enter. Sealing these gaps can reduce the amount of flow between the building and the outside. This not only improves energy efficiency, but it also slows the rate at which contamination enters the building from the outdoors.

Many large buildings have a small surface-to-volume ratio. In such buildings, weatherization is unlikely to have a large effect on reducing casualties from an outdoor chemical/biological release. On the other hand, this improvement might pay for itself in a few years and improve occupant comfort, especially in the offices on the perimeter of the building. In addition, since weatherization may potentially improve safety for a chemical/biological attack, this action is worth consideration.

CHEMICAL OR BIOLOGICAL ATTACK: INDOOR RELEASE

**Distinguishing Between a
Biological Release and a Chemical Release**
Very rarely will biological agents cause immediate symptoms. On the other hand, chemical agents will. In terms of a biological release, the objective is to reduce the total number of people exposed to the agent. In order to accomplish this, efforts must be made to identify all people exposed to the agent. In terms of a chemical release, the objective is to minimize the concentrations to which people are exposed.

Indoor Releases
For any indoor release, whether that release is chemical or biological, evacuation should be done. Safely evacuating building occupants to a meeting point upwind of the building is the objective.

Animation Showing How HVAC Spreads Contamination
To view an animation showing how the HVAC system spreads contamination through a part of the building, log onto *http://securebuildinggs.lbl.gov.*
The animation runs at faster than real-time and shows the spread of gas after a very rapid release of gas that is not from a continuous source.

Biological or Unknown Release
In the event of this kind of occurrence, the HVAC system should be shut down and outdoor air dampers should be closed. If this action is not possible, the dampers should be put into full recirculation mode. Those local exhausts that serve bathrooms and kitchens must also be shut off since they are often controlled separately from the HVAC system. By performing these actions, the operator can prevent the building from becoming a source of contamination for anyone outside the building.

If possible, stairwells should be pressurized with 100% outdoor air to provide an evacuation route. Other HVAC, bathroom and utility room fans should be shut off.

Any persons known to be exposed to a biological or unknown release should be separated to avoid contaminating others. Persons exposed should receive medical treatment and should be decontaminated.

Chemical Release

In the event a chemical release, it is best to leave the HVAC system operating unless a knowledgeable HVAC operator is available to perform HVAC manipulation. If the building operator has checked system operation and is sure that dampers and fans are working correctly, some HVAC manipulations can be beneficial. The simplest action is to put the building on 100% outside air—no recirculation—with supply and exhaust fans on full power.

Other actions that could be taken include:

- Shutting off the supply from any air intake in which the possibility of a release into the air intakes is thought to have occurred;

- Pressurizing stairwells with 100% outdoor air providing a safe evacuation route;

- Putting air handlers that serve heavily contaminated areas into full exhaust; supply to those areas should be shut off, forcing air to flow from safe areas to contaminated areas; and

- Providing 100% outdoor air to uncontaminated areas and occupied areas—as long as the chemical is not being released into the air intakes.

It is important to note that depending upon the design of the HVAC system, some of these actions may be performed by putting the building into a "smoke removal" mode.

For a release into a building air intake, the supply from that intake should be shut off. However, the recommended actions for an indoor release should be followed, as well. In particular, the building should be evacuated.

Open Air Outdoor Release

In the event of an open air outdoor release, the following should be performed:

- Shut off HVAC fans;

- Close fresh air intake dampers;

- Close exhaust dampers (exhaust dampers can act as intakes with the HVAC off);

- Turn off exhaust fans in bathrooms, utility rooms, kitchens, etc. (these fans are usually controlled separately from the HVAC system);

- Close windows and doors;

- Remain indoors unless an evacuation order is given; and

- Report to a designated safe zone in the interior of the building when remaining indoors.

Chemical or Biological Attack—Unknown Release Location

It would take a substantial or highly toxic outdoor release to affect a building's occupants. Should this occur, there would most likely be obvious signs—dead animals, people collapsing on the street, etc. An outdoor release of this magnitude would affect occupants in the building (because all ventilation zones get some outdoor air). An indoor release, on the other hand, will affect some areas of the building more quickly and more severely than other areas. If there are no visible signs of an outdoor release and if some areas of the building appear to be more severely affected than other areas, then it must be assumed that the release occurred indoors.

A release into one or more of the building's air intakes is also possible. Whether the release is in the air intake or indoors, the air supply should be shut off for any area that is known to be contaminated. That area then should be put on full exhaust.

Sources

National Institute for Occupational Safety and Health, *Guidance for Filtration and Air-Cleaning Systems to Protect Building Environments from Airborne Chemical, Biological, or Radiological Attacks*, DHHS (NIOSH) Publication No. 2003-136. April 2003.

Lawrence Berkeley National Laboratory, *Secure Buildings: Advice for Safeguarding Buildings Against Chemical or Biological Attack*, September 23, 2004.

Chapter 5

Dirty Bombs

BACKGROUND

Recent terrorist events have raised concern about the possibility of a terrorist attack that would involve the release of radioactive materials. The most common means of such an attack would, in all probability, involve the use of a "dirty bomb."

As defined by the United States Nuclear Regulatory Commission (NRC), a dirty bomb, or Radiological Dispersion Device (RDD), combines conventional explosives such as dynamite with radioactive materials in the solid, liquid or gaseous form. The bomb, itself, is intended to disperse radioactive material into a small, localized area around the explosion. This could possibly cause people and buildings to be exposed to the radioactive material contained in the bomb. The main purpose of a dirty bomb is to frighten people and to contaminate the targeted buildings or land. As such, the buildings or land would be rendered untenable for a long period of time.

It is believed that the conventional explosive, itself, would be more lethal than the radioactive material. At the levels created by most probable sources, not enough radiation would be present in a dirty bomb to kill people or to cause severe illness. For example, radioactive material is used in hospitals for the diagnosis or treatment of cancer. This radioactive material itself, is sufficiently benign so that approximately 100,000 patients a day are released from hospitals with this radioactive material in their bodies.

However, there are other radioactive materials that could cause problems. For example, these radioactive materials, once dispersed in the air could contaminate several city blocks. One of the objectives of the terrorist is to cause public fear and possible panic. The release of a dirty bomb and its contents would cause a disruption to a company's normal business operations as well as to incur costly cleanup efforts.

A second type of RDD might involve a powerful radioactive source hidden in a public place. A trash receptacle in a busy train or subway station, trash receptacles or decorative planters in front of buildings, any area or location where large numbers of people pass through or congregate are potential sources for the housing of a RDD. Any dispersal of radioactive material in these areas carries the potential for people who pass by or congregate in these areas to get a significant dose of radiation.

It is important to note that a dirty bomb is not similar to a nuclear weapon. As noted previously, the major impact of a dirty bomb is produced by the blast and the resulting panic and fear that will ensue. While they are explosive devices that are used to spread radioactive materials, RDDs are not particularly effective means for exposing large numbers of people to lethal doses of radiation. They are likely to affect relatively small areas. Because the materials will disperse as a result of the explosion, only the areas near the blast will be contaminated. The level or degree of contamination will depend upon how much radioactive materials were in the bomb, as well as the weather conditions at the time of the blast.

A nuclear detonation, on the other hand, is a catastrophic event. In addition to the concomitant nuclear fallout and associated damage to structures, such a detonation would severely disrupt civil authority and the infrastructure. Evacuation procedures would be both complicated and compromised, as would the resumption of normal operations within the country.

THE IMPACT OF DIRTY BOMBS

In addition to creating fear and requiring costly cleanup, the primary impact from a dirty bomb containing low-level radioactive sources would be the blast, itself. Promptly detecting the kind of radioactive material used in the bomb would assist local authorities in advising the affected population of what protective measures to take. Such measures would include leaving the immediate area, or going inside until further directed. Gauging how much radiation is present would be difficult to do when the sources of the radiation are unknown. However, there would not be enough radiation in a dirty bomb to cause severe illness from exposure to radiation at the levels created by most of these probable sources.

It is important to note, though, that certain radioactive materials dispersed in the air could conceivably contaminate several city blocks. The extent of the local contamination would depend on a number of variables. These variables include the size of the explosive device, the amount and type of radioactive materials used in the device, and the weather conditions at the time of the attack.

SOURCES OF RADIOACTIVE MATERIAL

There is wide-spread speculation regarding terrorist accessibility to the various radioactive materials that could be placed in a dirty bomb. The most harmful radioactive materials are found in nuclear power plants and nuclear weapons sites. However, increased security at these facilities makes obtaining such materials much more difficult.

Because of the danger and difficulty involved in obtaining high-level radioactive materials from a nuclear facility, there is a greater chance that the radioactive materials used in dirty bombs would come from low-level radioactive sources. Such low-level radioactive sources are found in hospitals, on construction sites and at food irradiation plants. The sources in these areas are used to diagnose and treat illnesses, sterilize equipment, inspect welding seams and irradiate food to kill harmful microbes.

For more information about radiation and emergency response, see the Centers for Disease Control and Prevention's website at: http://www.bt.cdc.gov or contact the following organizations:

- The CDC Public Response Source at 1-888-246-2675
- The Conference of Radiation Control Program Directors [http://www.crcpd.org/]
- The Environmental Protection Agency [http://www.epa.gov/radiation/rert/]
- The Nuclear Regulatory Commission [http://www.nrc.gov/]
- The Federal Emergency Management Agency (FEMA) [http://www.fema.gov/]
- The Radiation Emergency Assistance Center/Training Site [http://www.orau.gov/reacts/]
- The U.S. National Response Team [http://www.nrt.org/production/nrt/home.nsf]
- The U.S. Department of Energy (DOE) [http://www.energy.gov] at 1-800-dial-DOE

For more information on other radiation emergency topics, visit www.bt.cdc.gov/radiation or call the CDC public response hotline at (888) 246-2675 (English), (888) 246-2857 (Español), or (866) 874-2646 (TTY)

WHAT TO DO FOLLOWING A DIRTY BOMB EXPLOSION

Radiation cannot be seen, felt or tasted by humans. Subsequently, if people are present at the scene of an explosion, they will not know whether radioactive materials were involved at the time the explosion occurred. If people are not too severely injured by the blast and they are not directed to return to the building site, there is a series of steps that should be taken. According to the United States Department of Health and Human Services/Centers for Disease Control and Prevention (CDC), people who are present at the time of a dirty bomb blast should:

- Leave the immediate area on foot. People should not panic. Nor should they take public or private transportation such as buses, subways or cars. If radioactive materials were involved, taking public/private transportation could contaminate cars or the public transportation system.

- Go inside the nearest building. Doing so will reduce exposure to any radioactive materials that may be present at the scene.

- Remove clothing as soon as possible; clothing should be placed in a plastic bag and the bag sealed. By removing the clothing, most of the contamination caused by external exposure to radioactive materials will be removed.

- Save the contaminated clothing. This will allow for testing for exposure without invasive sampling.

- Take a shower or wash as best they can. Since washing reduces the amount of radioactive contamination on the body the effects of total exposure are effectively reduced.

- Be on the lookout for information. Once emergency personnel can assess the scene and the damage, they will be able to tell people whether radiation was involved.

Note that the CDC advises that by listening to television and/ or radio broadcasts, people will be told if radiation was involved. If radioactive materials were released, people would be told where to report for radiation monitoring and blood testing. This monitoring and blood testing would be used to determine whether people were exposed to the radiation, as well as what steps would need to be taken to protect their health. Being at the site where a dirty bomb explodes does not necessarily guarantee that people will be exposed to radioactive materials. Until doctors are able to check people's skin with sensitive radiation detection devices, it will be uncertain whether they were exposed. Doctors will be able to assess risks after the exposure level has been determined.

It is important to note that even if people who were present at the time of a dirty bomb blast do not know whether radioactive materials were present, following these steps would be helpful in reducing any possible injury from other chemicals that may have been present in the blast.

Sources

Department of Health and Human Services, Center for Disease Control and Prevention. CDC Fact Sheet, *Dirty Bombs*, July, 2003.

U.S. Nuclear Regulatory Commission, Office of Public Affairs. Fact Sheet, *Dirty Bombs*, March, 2003.

Chapter 6

Biological And Chemical Releases

Following the September 2001 terrorist attacks, a series of intentional biological agent releases made the fear of such releases, as well as chemical attacks, a reality. Using the United States mail as the delivery system, the anthrax-laced envelopes were sent to members of the United States Congress and other high profile individuals, including members of the media. As a result of these intentional and deliberate terrorist attacks, several deaths occurred, mail service was interrupted in various areas, and a widespread fear of handling delivered mail occurred. Insidious in nature, biological and chemical agents pose serious threats to the health and safety of the public.

BIOLOGICAL THREATS

Biological agents are organisms or toxins that can kill or incapacitate people, livestock and crops. The three basic groups of biological agents which would likely be used as weapons are bacteria, viruses and toxins.

Bacteria are small, free-living organisms that reproduce by simple division and are easy to "grow." The diseases they produce often respond to treatment with antibiotics.

Viruses are organisms which require living cells in which to reproduce and are intimately dependent upon the body they infect. Viruses produce diseases which generally do not respond to antibiotics. However, antiviral drugs are sometimes effective.

Toxins are poisonous substances found in and extracted from living plants, animals and/or microorganisms. Some toxins can be produced or altered by chemical means. Antitoxins and selected drugs can be used in the treatment of some toxins.

Most biological agents are difficult to grow and maintain. Many biological agents can break down quickly when they are exposed to sunlight and other environmental factors. Other biological agents, like anthrax spores, are very long lived. They can be dispersed in several ways, including: spraying them in the air; contaminating water and food; and/or infecting animals, which can carry the disease to humans.

When dispersed in the air as an aerosol, the biological agent forms a fine mist that can drift for miles. Inhaling the agent can cause disease in people or animals. While most microbes can be destroyed and toxins deactivated by cooking food and boiling water, some organisms and toxins can persist in the food and water supplies. If infected, insects and animals such as fleas, mice, flies and mosquitoes can spread diseases to humans.

Individuals can spread infectious agents to other people, as well. For example, humans have been the source of infection for smallpox, plague and the Lassa viruses.

CHEMICAL THREATS

Chemical agents are poisonous vapors, aerosols, liquids or solids that have toxic effects on people, animals or plant life. They can be released by bombs, sprayed from aircraft, boats or vehicles. They can also be used as a liquid to create a hazard to people and the environment. Some chemical agents may be odorless and tasteless. Their effect can be immediate— from a few seconds to a few minutes; or there can be a delayed effect with the onset of symptoms or illness occurring several hours to several days after exposure.

While potentially lethal, chemical agents are difficult to deliver in fatal concentrations. Outdoors chemical agents often dissipate quickly. They are also difficult to produce.

There are six types of chemical agents:

(1)   Pulmonary, or lung-damaging agents such as phosgene;

(2)   Cyanide;

(3)   Vesicants, or blister agents such as mustard;

(4)   Nerve agents such as GA (tabun), GB (sarin), GD (soman), FG and VX;

(5)   Incapacitating agents such as BZ; and

(6)   Riot-control agents (similar to Mace®)

BIOLOGICAL ATTACKS

In many biological attacks, people will not know that they have been exposed to an agent. In such situations, the first evidence of an attack may be when a person notices symptoms of the disease caused by exposure to an agent. If this should occur, the person should seek immediate medical attention.

In some situations like the anthrax-laced letters that were sent in Fall 2001, people may be alerted to potential exposure. If this should occur, close attention to all official instructions must be heeded. Depending upon the nature and severity of the biological attack, the delivery of medical services for a biological occurrence may be handled differently.

If skin or clothing comes in contact with any visible or potentially infectious substance, the clothing should be removed and bagged. The skin should be immediately washed with warm soapy water, and medical assistance should be sought.

CHEMICAL ATTACKS

Some of the most immediate symptoms of exposure to chemical agents include blurred vision, eye irritation, difficulty breathing and nausea. As in the case of a biological attack, exposure to a chemical agent requires immediate medical attention. If such medical attention is not immediately available, persons exposed to a chemical agent must decontaminate themselves and assist in decontaminating others, within minutes of exposure to minimize health consequences. It is important to note that a person should not leave the safety of a shelter to go outdoors to assist others, until authorities have announced that it is safe to do so.

The decontamination procedures, as outlined by the Federal Emergency Management Agency (FEMA), include:

- Removing items in contact with the body, including jewelry (rings, watches, etc.) and other items such as hair clips. Contaminated clothing normally removed over the head should be cut off to avoid contact with the eyes, nose and mouth. The clothing should be placed in a plastic bag, if possible. Hands can be decontaminated by using soap and water. Eyeglasses or contact lenses should be removed and glasses placed in a pan of household bleach;

- Removing any and all items in contact with the body;

- Flushing eyes with a lot of water;

- Gently washing the face and hair with soap and water and then thoroughly rinsing with clear water;

- Decontaminating all other body areas that have been contaminated. Affected body areas should be blotted (never swabbing, scraping or scrubbing) with a cloth soaked in soapy water and rinsed with clear water;

- Changing into uncontaminated clothes. Clothing stored in drawers or closets is likely to be uncontaminated; and

- Proceeding to a medical facility for screening, as soon as possible.

NUCLEAR AND RADIOLOGICAL ATTACKS

Nuclear explosions can cause deadly effects: blinding light, intense heat (thermal radiation), initial nuclear radiation, blast, fires started by the heat pulse, as well as secondary fires caused by the destruction. Surface level explosions also produce radioactive particles called fallout that can be carried by wind for hundreds of miles.

Terrorist use of a radiological dispersion device (RDD), commonly referred to as a "dirty nuke" or "dirty bomb," is considered far more likely than terrorist use of a nuclear device. RDDs are a combination of conventional explosives and radioactive material designed to scatter dangerous and sub-lethal amounts of radioactive material over a given area. According to FEMA, the RDD appeals to terrorists for a number of reasons. First, the RDD requires very little technical knowledge to build and deploy, compared to the knowledge required to build and deploy a nuclear device. Secondly, the various radioactive materials used in RDDs are also widely used in medicine, agriculture, industry and research. As such, they are more readily available and easier to obtain, compared to the weapons grade uranium or plutonium that is used in nuclear devices.

Terrorist use of a nuclear device would probably be limited to a single smaller "suitcase" weapon. The strength of such a weapon would be in the range of bombs used during Work War II. The nature of the

Homeland Security Advisory System
The Homeland Security Advisory System was designed to provide a national framework and comprehensive means to disseminate information regarding the risk of terrorist acts to federal, state and local authorities and to the American people. This system provides warnings in the form of a set of graduated "threat conditions" that increase as the risk of the threat increases. At each threat condition, government entities and the private sector, including businesses and schools, would implement a corresponding set of "protective measures" to further reduce vulnerability or increase response capability during a period of heightened alert.

HOMELAND
SECURITY
ADVISORY SYSTEM

SEVERE (Red)

HIGH (Orange)

ELEVATED (Yellow)

GUARDED (Blue)

LOW (Green)

Although the Homeland Security Advisory System is binding on the executive branch, it is voluntary to other levels of government and the private sector. There are five threat conditions, each identified by a description and corresponding color.

The greater the risk of a terrorist attack, the higher the threat condition. Risk includes both the probability of an attack occurring and its potential gravity.

Figure 6-1. Homeland Security Advisory System

effects would be the same as a weapon delivered by an inter-continental missile, but the area and severity of the effects would be significantly more limited.

There is no way of knowing how much warning time there would be before an attack by a terrorist using a nuclear or radiological weapon. A surprise attack always remains a possibility.

Intentional Releases

An intentional biological or chemical release is predicated upon several assumptions: According to the Lawrence Berkeley National Laboratory, these assumptions are:

- the agent would be released into the air quickly;
- the agent would be virulent;
- the release would be unexpected;
- the location of the release would be unknown; and
- the total mass of the agent released into the air would be less than 10 kg.

Typically, large commercial buildings have fairly ordinary HVAC systems. These buildings contain multiple air handling units. These units supply air to separate areas of the building and are designed to draw return air from the same area that they supply. Generally speaking, few HVAC systems work as designed. In many cases, some air is carried between ventilation zones by pressure imbalances, including the stack effect and wind effects.

Further, in some large buildings each air handling unit supplies air to a different part of the building, but the return air enters a "common return" and is a mixture of air from several of these areas. In these cases, the HVAC system can spread contamination as if there were a single air handling unit serving all of the areas.

The Nature of Biological and Chemical Releases

Once it has been determined that a release of some kind has occurred, it is important to determine the type of release—biological or chemical. The release of a biological agent rarely causes immediate symptoms while a chemical agent can produce almost immediate symptoms.

Exposure to toxic chemicals, including chemical warfare agents, display one or more of the following symptoms:

- Pinpoint pupils, leading to a perception of darkness
- Dilated pupils (note that dilated pupils can be caused by some chemicals, but not chemical warfare agents)
- Dizziness
- Runny nose
- Clammy skin/perspiration
- Breathing difficulties
- Nausea and/or vomiting
- Blurred vision or blindness
- Seizures
- Loss of bladder control
- Loss of consciousness or death

If building occupants exhibit sudden onset of some of these symptoms, a chemical release may be responsible. Although food poisoning displays many of these symptoms, food poisoning, as such, would not strike as many people simultaneously.

DISTINGUISHING BETWEEN INDOOR AND OUTDOOR RELEASES

The location of the source is critical, since the responses will differ for the outdoor release, the indoor release and a release into the building air intakes.

Biological releases are difficult to determine; the symptoms of a biological release are not immediately apparent. Unless the release was directly observed, or a warning was issued by an official agency, detecting the source of a biological release may not be possible.

On the other hand, determining the location of a chemical release may be somewhat easier. If the release occurs near the building's air intake, building occupants would experience the earlier described symptoms. If these symptoms are not experienced by building occupants, an outdoor release would be indicated. Visually inspecting the immediate outdoor area from the confines of the building, itself, would confirm an outdoor release. This visual evidence would be dead or dying birds and other wildlife, people collapsing on the sidewalks, etc.

Outdoor Releases

Regardless of a biological or chemical occurrence, the following actions should be taken to minimize exposure in the event of an outdoor release:

- Keep people indoors

- Close all windows and doors to the outside

- Close all internal doors

- Shut of all HVAC fans and close all HVAC dampers, including exhaust dampers

- Shut off other fans such as kitchen and bathroom exhausts

- Do not use elevators (they can pump air into or out of a building)

- Have occupants assemble in pre-identified shelter-in-place rooms (rooms with no, or low air exchange with the outdoors, and which have low air exchange with the rest of the building)

- Once the emergency response teams have determined that the outdoor concentration has diminished to safe levels, evacuate the building and flush the building with outdoor air. Since buildings tend to retain contamination that has entered, the concentration of contamination will be higher inside the building than outside.

If buildings are equipped with special filtration systems, it may be beneficial to continue HVAC operations. Minimizing the rate of air exchange with the outside will keep the indoor concentration as low as possible for as long as possible. Normal operation of HVAC will exhaust some building air and pull in some outdoor air. If the outdoor air is contaminated, the HVAC system will spread the contamination throughout the building. Air exhausted from the building by exhaust fans will also be replaced by outdoor air. Shutting off HVAC fans and exhaust fans will help minimize the air exchange with the outside.

Even putting the system on full recirculation, if that is possible, is generally not as good as shutting off the HVAC, since duct systems and dampers normally allow substantial leakage.

Special filters that can provide protection against a chemical release are also available. Expensive, they are usually installed in those buildings that are or have been determined to be at high risk. When these filters are installed and the building positively pressurized with respect to the outdoors, continued HVAC operations may be beneficial in the event of an outdoor chemical release.

Indoor Release: Biological or Chemical

When a release occurs through the building's air intakes, the biological or chemical agent inside the building presents a very clear danger to building occupants. Because the concentration of the chemical agent will be thousands of times higher inside the building than outside, evacuation procedures should be initiated. Once occupants are outside and upwind of the building, exposure to the airborne agent will stop. However, exposure to the agent, itself, from contaminated clothing and skin may continue.

Evacuation poses several risks in and of itself. Evacuation often requires occupants to use stairways and hallways which tend to be more contaminated than closed offices. This is true especially if HVAC operations have not been performed, or poorly performed. Once occupants are evacuated outside and upwind of the building, a secondary attack could occur such as a car bomb targeting the evacuees. Also, in a major metropolitan area, passers-by could be contaminated unknowingly, posing a direct risk to the passers-by. This could potentially increase the chance of starting an epidemic if the disease is contagious.

Where to Assemble

According to the Lawrence Berkeley National Laboratory, occupant evacuees should congregate at one or two meeting points upwind of the building and at least 100 feet or farther from the building, itself. It is important to separate the exposed occupant evacuees from those not exposed as soon as possible.

Another possible assembly area, where feasible, would be to have occupant evacuees congregate by floor. Assembling occupant evacuees at predetermined assembly areas assures that occupants can be accounted for, thereby eliminating needless searches into the building. Predetermined assembly areas can also be used to provide first aid or treatment for any chemical or biological agent exposure. Predetermined assembly areas also separate exposed occupant evacuees from passers-by.

Indoor Release—Biological

In many cases, a biological agent will not cause immediate symptoms. In fact, the type of biological agent may not be known for several days after the release. The insidious nature of biological agents rests with the fact that they can be introduced into a building without the occupants' knowledge.

However, if it has been determined or suspected that a biological agent has been introduced into a building, there are a number of actions that must be undertaken. These actions include: limiting the number of occupants exposed; closing HVAC dampers and turning off fans; pressurizing stairwells with outside air; and separating exposed occupants from other building occupants.

Limiting Exposure

As previously discussed, limiting the number of people exposed to a biological agent is the goal. This is important for obvious reasons:

1. If the agent-producing disease is treatable, the fewer people exposed will result in ensuring that everyone needing treatment will receive it.

2. Conversely, if the agent-producing disease is untreatable, the fewer persons exposed will result in fewer casualties.

3. If the agent-producing disease has the potential to cause an epidemic (i.e., the disease is contagious), then having the capability to identify everyone exposed is critical.

Protective Procedures—Biological

The following procedures are recommended for a biological release:

*   Close HVAC dampers and turn off all fans;
*   Pressurize stairwells with outdoor air; and
*   Isolate exposed occupants.

Close HVAC Dampers and Turn off All Fans

HVAC systems operation will exhaust air and the biological agent to the outdoors. As a result, the HVAC system can transmit the infectious agents to not only building occupants but also to individuals outside the building. While building occupants may feel thermal discomfort while the HVAC system is shut off—either too hot or too cold—there is no risk from oxygen deprivation or from carbon dioxide build-up for many hours/duration of the shut down.

On the other hand, with the HVAC system shut down, the change in airflow patterns may allow air to flow between ventilation zones that would otherwise remain partially isolated. As such, more occupants would be potentially exposed to the biological agent.

Therefore, it is important to note that whatever HVAC actions are taken, everyone in the building should be considered as potentially exposed. While the objective is to minimize the number of people exposed, the key point is to isolate everyone who has been exposed so that medical treatment can be provided.

Pressurize Stairwells with Outdoor Air

Pressurizing stairwells with 100% outdoor air provides a safe evacuation route. This pressurization prevents contaminated air from being pulled into the stairwells, as can happen often due to the "stack effect." All other ventilation vents should be shut off, so that the building does not become a significant conduit for the release of the biological agent.

Isolate Exposed Occupants

Occupants, who are known to be exposed to the biological agent, should be isolated. Isolating those individuals prevents contamination of others through contact with clothing or skin. Isolation of exposed individuals expedites medical treatment and decontamination.

Indoor Release—Chemical

There are several recommended actions that should be carried out for an indoor release of a chemical agent. They are:

1. Minimize exposure
2. Choose the appropriate HVAC operation:
   a. Continue HVAC operation; default action
   b. Set HVAC to provide outdoor air per the authorized HVAC operator
   c. Best action: Perform sophisticated HVAC manipulation

Minimize Exposure

Unlike a biological release, the goal for addressing a chemical release is to minimize occupant exposure. Minimizing exposure for building occupants is done by exhausting contaminated air from the building and replacing it with outdoor air. Once contaminated air is expelled from the building, it rapidly dilutes and becomes less harmful. Because the contaminated air that is expelled will be replaced by uncontaminated air, exposure is reduced for occupants still in the building. It is important to note, however, that some contaminated air may reenter the building. The variables that may cause reentry are the location of the building air intakes relative to the exhausts, prevailing wind direction, and speed.

Figures 6-2 through 6-3, "Information for First Responders," provide basic information to facility managers and building owners about HVAC design and operation for typical commercial buildings.

Figures 6-4 and 6-5, "Advice for Building Operators and Incident Commanders," illustrate actions rather than general building knowledge.

Choosing the Appropriate HVAC Operation

Choosing the appropriate HVAC operation involves several actions including: continuing HVAC operation; setting the HVAC system for outdoor air; and performing sophisticated HVAC manipulation.

Continuing HVAC Operation

There are two schools of thought on continuing HVAC operations. The Lawrence Berkeley National Laboratory suggests that it is best to leave the HVAC system operating without alteration, unless a knowledgeable building operator is available to perform HVAC manipulations. On the other hand, the U.S. Army Corps of Engineers suggests that all air-handling units be shut down until the type of hazard and extent of the release can be determined.

In either case, it is important to note that in buildings (particularly those taller than a few stories) are subject to the "stack effect." Simply put, if the indoor air is cooler than outdoor air, building air will tend to flow out through the bottom levels of the building and be replaced by air coming in the top. Correspondingly, the reverse happens if the indoor air is warmer than the outdoor air. With the stack effect, the vertical flows may draw contaminated air into the vertical connections between the floors. This occurs most notably in elevator shafts and in stairwells which are the main evacuation routes. In very tall buildings, especially during the winter season, the stack effect can create pressure differences so large that elevator doors will not function properly, stairway doors can be difficult to open, and airflows in elevator shafts and stairwells can be substantial. These effects are strongly dependent on air leaks in the building's upper floors. These can be exacerbated if occupants open/break windows on the upper levels of the building in an attempt to ventilate without outdoor air.

First Responders to an Indoor Chemical Release
Ventilation System ON

1. If supply air becomes contaminated, contaminant will spread rapidly through the entire ventilation zone: every supply register in that zone becomes a source. This can happen:

   (a) if contaminated outdoor air enters the intake.

   (b) if contaminant-bearing air from inside the building is recirculated.

2. Most commercial buildings recirculate indoor air, if outdoor air is hot or cold. In this case, supply air (into the building) will eventually become contaminated, so the pollutant will spread everywhere.

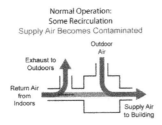

(a) Most ventilation systems supply a mix of outside air and recirculated (return) air.

(b) In extreme hot and cold weather, the mix shifts to higher recirculation.

(c) In mild temperatures (55-70°F), some buildings take in as much outdoor air as possible.

3. The ventilation system causes large air flows that move contamination through the building.

   A ventilation zone may cover a large or small area, and may mix air between floors.

   (a) A ceiling air return plenum may serve a single room or a large zone; contaminated air can be pulled along in the plenum and may quickly enter the supply air (1-10 min.)

   (b) If a contaminated room has a "ducted return", contaminated air will probably enter the supply air very rapidly (20 sec - 3 min).

   (c) If supply air is contaminated, contamination will spread throughout the whole ventilation zone rapidly (seconds or minutes): Every supply register in the zone becomes a source.

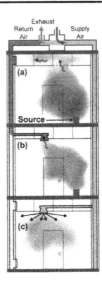

Source: Lawrence Berkeley National Lab

Figure 6-2. Information for First Responders to an Indoor Chemical Release—Ventilation System ON.

Information for
First Responders to an Indoor Chemical Release
Ventilation System ON

4. A return air grille in a hallway can draw contaminated air into and along the hall, even if doors are closed.

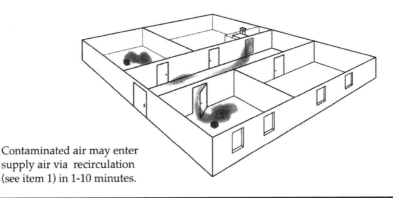

Contaminated air may enter supply air via recirculation (see item 1) in 1-10 minutes.

5. A stairwell, elevator shaft, or utility chase can provide a pathway for flow between floors. The ventilation system can force airborne contaminants to flow either up or down.

A moving elevator creates a piston effect that can force contaminatation to flow up or down.

Flows can be significant even if elevator doors and stairwell doors are closed.

Unlike smoke, contaminants can be either heavier or lighter than air, and so can sink or rise even in still air.

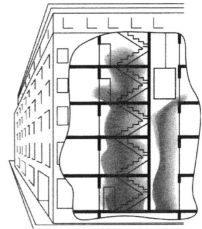

Source: Lawrence Berkeley National Lab

Figure 6-2. Information for First Responders to an Indoor Chemical Release—Ventilation System ON (*Cont'd*).

Information for

First Responders to an Indoor Chemical Release
Ventilation System OFF

1. Effects that can be ignored when the ventilation system is on, become dominant when it's off. Examples are wind leaking into the building, drafts, and buoyancy (warm air rises, cool air sinks).

2. Air flows are generally slower than when the ventilation system is on. Ventilation ducts provide pathways for contamination to flow between rooms and floors, even with the ventilation system turned off.

Temperature and pressure differences can drive flow upward or downward between floors. Contaminant can flow from room to room, for example:

(a) horizontally through ducts

(b) vertically through ducts or other openings

(c) through the ceiling plenum

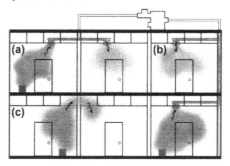

3. Flows depend strongly on wind and on the indoor-outdoor temperature difference, especially when windows are open.

Outdoors Warmer:
Indoor air, which carries the contamination, tends to descend as it leaks from the building. Outside air enters upper floors. Some contamination may still move upwards due to local flows or drafts.

Outdoors Cooler:
Indoor air, which carries the contamination, tends to rise as it leaks from the building. Outside air enters lower floors. Some contamination may still move downwards due to local flows or drafts.

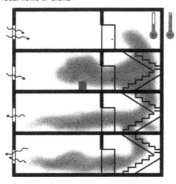
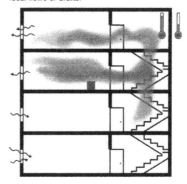

Source: Lawrence Berkeley National Lab

Figure 6-3. Information for First Responders to an Indoor Chemical Release—Ventilation System OFF.

Information for

First Responders to an Indoor Chemical Release Ventilation System OFF

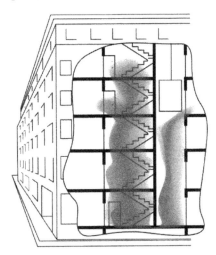

4. Fast or strong vertical flows can occur through elevator shafts, stairwells, utility chases, and other connections between floors.

Horizontal flows are usually weaker than vertical flows, except when there are strong winds or other causes of horizontal pressure differences.

Unlike smoke, contaminants can be either heavier or lighter than air, so they can sink or rise even in still air.

Source: Lawrence Berkeley National Lab

Figure 6-3. Information for First Responders to an Indoor Chemical Release—Ventilation System OFF (*Cont'd*).

Setting the HVAC System to Provide Outdoor Air.

If the HVAC operator has previously checked system operation, and is sure that dampers and fans are working correctly, the Lawrence Berkeley National Laboratory suggests that the HVAC system can be set to supply 100% outdoor air. By doing so, the chemical agent will be diluted and exhausted from the building without spreading it through the HVAC system. This action is warranted because it is better than an unchanged HVAC action which typically recirculates contaminated air.

Performing Sophisticated HVAC Manipulation

There are several beneficial HVAC manipulations that can be performed by a knowledgeable operator. This person should have a good working knowledge about the building's operation including knowledge of which air handlers serve which ventilation zones and how to control dampers to close off supply to some areas. These actions have the potential to reduce chemical exposure of occupants in building areas

Advice for
Building Operators and Incident Commanders
Response to an Indoor Chemical Attack

Unless a knowledgeable building operator is present:

- Leave the HVAC system operating as is.

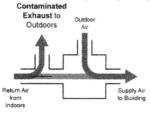

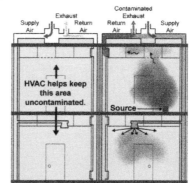

Under normal operation, the HVAC system will exhaust contaminated air from the building, and replace it with fresh air.

WARNING: a plume of contamination will spread downwind from the building's exhaust vents.

Continued HVAC operation may slow chemical spread between areas served by different air handling units, and help prevent contamination of stairways and hallways.

If a knowledgeable building operator is present:

- Set fans and dampers to deliver 100% outdoor air at maximum volume (see note 1 below).
- If a release into one or more of the building's air intakes is suspected, shut off supply from the contaminated air intakes.

If more sophisticated actions are possible:

- Pressurize stairwells with 100% outdoor air (see note 2).
- Put the air handlers serving heavily contaminated areas onto full exhaust (see note 3).
- Shut off supply to contaminated areas (see note 3).
- Provide 100% outdoor air to uncontaminated areas and areas with people.

Notes:

1) Delivering 100% outdoor air will provide safe air to occupants and will exhaust the chemical quickly.

2) Pressurizing stairwells with 100% outdoor air will help provide a safe evacuation route.

3) Exhausting contaminated areas and supplying fresh air to uncontaminated areas helps ensure that air does not flow from contaminated areas to safe areas.

4) Depending on the HVAC design, some of the more sophisticated actions may be achieved by putting the building into "smoke removal" mode.

If safely possible, evacuate people from the building to a meeting point upwind of the building.
Visit **http://securebuildings.lbl.gov** for more information and updates.

Source: Lawrence Berkeley National Lab

Figure 6-4. Advice for Building Operators and Incident Commanders—Response to an Indoor Chemical Attack.

Advice for

Building Operators and Incident Commanders
Response to an Indoor Biological Attack

It is critical to find and treat everyone who has been exposed.

To help prevent exposing people outside the building:

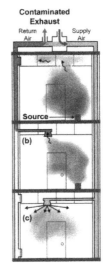

- Shut off the HVAC system.

- Close intake and exhaust dampers (or, if this is not possible, set them for full recirculation). Leave HVAC off.

- Shut off local exhausts, such as those serving bathrooms and kitchens. They are often controlled separately from the HVAC system.

To help reduce exposure of building occupants:

- Pressurize stairwells with 100% outdoor air if possible, to provide an evacuation route.

- Segregate people known to be exposed, to avoid contaminating others via contact with clothes or skin, and tag or mark these people for medical treatment and decontamination.

Normal HVAC operation releases contaminated air to the outdoors.

- If possible, evacuate people from the building to a meeting point upwind of the building.

Notes:

1) Pressurizing stairwells with fresh air will help keep contaminated air from entering the stairwells.

2) Everyone exposed should receive treatment, as symptoms may not appear for several days.

Visit **http://securebuildings.lbl.gov** for more information and updates.

Source: Lawrence Berkeley National Lab

Figure 6-5. Advice for Building Operators and Incident Commanders— Response to an Indoor Biological Attack.

that have not yet been contaminated.

A knowledgeable HVAC operator should perform the following actions:

1. The operator should be aware of the possibility of a source in the building air intakes, and should shut off the supply from any intake in which this is thought to have occurred;

2. Pressurize stairwells with 100% outdoor air;

3. Put the air handlers that serve heavily contaminated areas onto full exhaust and shut off supply to those areas; and

4. Supply 100% outdoor air to uncontaminated areas and areas with people in them.

For those buildings that have a smoke control or a smoke removal system it is important to note that the system's effectiveness lies in the fact that smoke removal systems often have their own duct-work and exhaust registers, rather than using the HVAC system. As such the smoke system's exhaust ducts act as a conduit to the outdoors.

It is important to note that a chemical release can be removed effectively using a smoke control system. However, chemicals and chemical agents differ from smoke. Chemicals and chemical agents are not buoyant; smoke is. Smoke infiltration barriers that rely on buoyancy will not be effective. Also, staying low to the floor will not generally reduce exposure.

By pressuring stairwells with 100% outside air, a safe evacuation route is provided. In a large building evacuation, the amount of time spent in the stairwells is likely to be high. Most everyone will need to use a stairwell and, as such, they may spend more time in the stairwell than in other areas of the building. Many buildings already have equipment that can pressurize stairwells with outdoor air since this is a standard fire safety technique.

Putting air handlers onto full exhaust and shutting off supply to those areas will de-pressurize the affected areas relative to the rest of the building. Air is forced to flow from safe areas to contaminated areas

Even with air handlers that serve contaminated areas set to full exhaust and the supply dampers closed, uncontaminated air will still be delivered to the contaminated areas. These areas will draw air from adjacent uncontaminated areas and from the outside and through the building shell. While the supply of air will be less than if these areas received supply air by the HVAC system, the fact remains that uncontaminated air will be delivered.

By providing 100% outdoor air to uncontaminated occupied areas ensures that occupants in these areas remain safe particularly if this action is coupled with putting the contaminated areas on full exhaust.

Sources

Lawrence Berkeley National Laboratory, *Protecting Buildings From a Biological or Chemical Attack: actions to take before or during a release.* LBNL/PUB-51959. January, 2003.

U.S. Army Edgewood Chemical Biological Center and the Army Corps of Engineers Protective Design Center, *Technical Instruction 853-01. Protecting Buildings and Their Occupants from Airborne Hazards,* October, 2001.

Chapter 7

Regulatory Compliance

ollowing the September 11th attacks, the definition of building hazards in the United States has changed to include intentional attack. Protection of the population from acts of terrorism has become a major national priority.

Though historically focused on fire safety, the building regulatory system does address natural disaster mitigation (e.g., floods, wind/snowstorms, earthquakes, etc.), some human engineered risks (e.g., HAZMAT storage), and specific federal mandates (EPA, ADA, etc.). The regulation of these areas is supported by well-established and accepted reference standards, regulations, inspection and assessment techniques, planned review methods, and quality control. The areas that are addressed are zoning and planning regulations, property maintenance codes, building rehabilitation codes and construction codes.

BUILDING REGULATION ENFORCEMENT

- **Zoning and planning regulations** define land use, building density, transportation systems, and utility systems. While usually adopted by local and municipal governments, various state agencies may provide guidance and input. For existing vulnerable properties, these regulations can address specific access-control measures. This requires prioritization of hazards and buildings.

- **Property maintenance codes** govern the use and maintenance of existing buildings. Housing and fire codes are examples of property maintenance codes. They are developed by model code and consensus standards organizations and adopted as regulations by local government agencies. These codes are effective at addressing building vulnerabilities. Buildings thus require extensive inspections to ensure compliance.

- **Building rehabilitation codes** address health, safety, and welfare in existing buildings that are undergoing voluntary improvement. These codes can be effective at addressing vulnerabilities in existing buildings. A relatively new development, building rehabilitation codes have been enacted by some state and/or local government agencies.

- **Construction codes** include building, mechanical, plumbing and electrical standards. They address the health, safety and welfare issues in new building construction. They are developed by model code organizations (e.g., the National Fire Protection Association, NFPA) and adopted as regulations by state and/or local government agencies. In most cases, the Federal Government does not regulate construction requirements. However, a notable exception is the Americans with Disabilities Act (ADA). The ADA is enforced by the U.S. Department of Justice. Construction codes can be effective at addressing the problem of vulnerability to risk.

CURRENT REGULATORY STATUS

Current building codes address and are effective at mitigating the effects of fire and natural disasters. Codes also regulate indoor air quality as well as the installation of mechanical, plumbing, electrical and communication systems.

Codes and the Risk of Terrorism

- *Bomb blasts* are not addressed in codes, but some of the earthquake and windstorm provisions in building codes may have a beneficial effect on mitigating the effects of a bomb blast. Code-regulated earthquake design requires the building's structural system to withstand pressure from the force of ground shifts, helping to mitigate the effects of bomb blasts. Helping to mitigate the hazards of flying glass in explosions are the code-regulated hurricane design which requires windows and glass to resist the effects of the impact of windborne debris.

- *Chemical, biological, and radiological agents* are not addressed in building codes. There are, however, certain details in the design of building heating, ventilation, and air conditioning (HVAC) systems that are regulated by mechanical codes. These codes may mitigate the effects of these chemical, biological, and radiological agents.

* *Progressive collapse* is one of the effects of a blast although it is not the only one. It is discussed in the American Society of Civil Engineers standard, ASCE 7. This is the structural loads standard that is referenced in building codes. While no design criteria are specified, some qualitative guidance is provided. ASCE 7 and the American Concrete Institute standard, ACI 318 (N.B.: ACI 318 is the reinforced concrete design standard reference in building codes) have references to structural integrity but not as a set of criteria for resisting progressive collapse.

* In cases of *Armed attack* the codes regulating the design and construction of correctional facilities may be helpful in the protection of new and/or existing buildings.

REGULATORY ACTIVITIES
RELATED TO THE RISK OF TERRORISM

Occupational Safety and Health Administration (OSHA)
Emergency response activities as a result of a bioterrorist attack are addressed in specific standards for general industry. Note that some states have OSHA-approved state plans and have adopted their own standards and enforcement policies. As in all cases, the most stringent standard prevails. Building owners and facility managers should contact their respective state occupational safety and health offices for further information.
Specifically the OSHA standards that address bioterrorism include:

* **1910.120.** This is the entire HAZWOPER standard. It addresses the issue of hazardous waste operations and emergency response. **1910.120(q)** of the standard addresses emergency response. The directive, Inspection Procedures for the Hazardous Waste Operations and Emergency Response Standard, **29 CFR 1910.120 and 1926.65, Paragraph (q)**: Emergency Response to Hazardous Substance Releases. **OSHA CPL 2-2.59A** (April 24, 1998), provides instruction

that establishes policies and provides clarification to ensure uniform enforcement of paragraph (q) of the HAZWOPER standard, **29 CFR 1910.120** and **1926.65.**

- **OSHA CPL 2-1.037**(July 9, 2002) details the compliance policy for emergency action plans and fire prevention plans.

OSHA standard 1910.38 which addresses the issue of emergency action plans is very specific in terms of employer requirements. Note the sections below:

- **1910.38(a)—Application.** Employers must have an emergency action plan.

- **1910.38(b)—Written and oral emergency action plans.** Emergency action plans must be in writing, kept in the workplace, and available to employees for review. However, employers with 10 or fewer employees may communicate the emergency action plan orally to employees.

- **1910.38(c)—Minimum elements of an emergency action plan.** Emergency action plans must include, at a minimum, the following:
  - 1910.38(c)(1)—Procedures for reporting a fire or other emergency;
  - 1910.38(c)(2)—Procedures for emergency evacuation including type of evacuation and exit route assignments;
  - 1910.38(c)(3)—Procedures to be followed by employees who remain to operate critical plant operations before they evacuate;
  - 1910.38(c)(4)—Procedures to account for all employees after evacuation;
  - 1910.38(c)(5)—Procedures to be followed by employees performing rescue or medical duties; and
  - 1910.38(c)(6)—The name or job title of every employee who may be contacted by employees who need more information about the plan or an explanation of their duties under the plan.

- **1910.38(d)—Employee alarm system.** Employers must have and must maintain employee alarm systems. The employee alarm system must use a distinctive system for each purpose and comply with the requirements in § 1910.165.

- **1910.38(e)—Training**. Employers must designate and train employees to assist in the safe and orderly evacuation of other employees.

- **1910.38(f)—Review of emergency action plan**. Employers must review emergency action plans with each employee covered by the plan:
  — 1910.38(f)(1)—When the plan is developed or the employee is assigned initially to a job;
  — 1910.38(f)(2)—When the employee's responsibilities under the plan change; and
  — 1910.38(**f)(3)**—When the plan is changed.

The Federal Emergency Management Agency (FEMA) has published *The World Trade Center Performance Study* "to examine the damage caused by these events, collect data… and identify studies that should be performed."

The New York City Department of Buildings, soon after the World Trade Center attack, initiated an effort to analyze the code as it relates to terrorist threat. In February 2003, the task force issued a report of findings and 21 specific recommendations for code, code administration, and code enforcement changes.

The American National Standards Institute (ANSI) established a Homeland Security Standards Panel (HSSP) in February 2003, in response to The National Strategy for Homeland Security. The proposed mission of the HSSP is to catalogue, promote, accelerate and coordinate the timely development of consensus standards within the national and international voluntary standards system. These standards are intended to meet identified homeland security needs. Another goal is to communicate the existence of such standards to governmental units and the private sector.

The National Fire Protection Association (NFPA), a standards organization active in the field of fire safety, established a committee on Premises Security before 9/11. It plans to produce 2 standards: NFPA 730, *Guide to Premises Security* and NFPA 731, *Security System Installation Standard*.

The **American Society for Testing and Materials** (ASTM), a standards organization active in the field of materials, specifications, and test methods, many of which are referenced in building codes, is considering the creation of a Homeland Security Committee, or subcommittee.

The **American Society of Heating, Refrigerating and Air-conditioning Engineers** (ASHRAE), a standards organization active in the field of mechanical systems and indoor air quality in buildings, may initiate activities addressing chemical, biological and chemical agents in buildings.

The **American Society of Mechanical Engineers** (ASME), a standards organization active in the field of boilers and pressure vessels, elevators, and other building equipment, has developed a program of seminars for engineers. These seminars entitled "Strategic Response to Terrorism," cover a range of topics including chemical and biological terrorist attacks.

The **National Institute of Standards and Technology** (NIST) is conducting an intensive review and analysis of the World Trade Center collapse. It is anticipated that this analysis will lead to code changes related to structural safety and fire safety in high-rise buildings.

The **General Services Administration** (GSA) published PBS-PQ100.1, *Facilities Standards for the Public Buildings Service*. It contains building design criteria for blast resistance, progressive collapse, and chemical, biological and radiological attack.

The **Department of Defense** (DoD), has a similar standard: *Unified Facilities Criteria, DoD Minimum Antiterrorism Standards for Buildings*, UFC 4-010-01, 31 July, 2002.

BUILDING REGULATION MANAGEMENT MODELS

The four categories of building regulations are zoning, property maintenance, building rehabilitation and building construction. Each has the potential to address the physical aspects of terrorism risk, including common terrorist tactics and delivery systems, as well as terrorist attack devices. Table 7-1 highlights the applicability of building regulation to terrorist tactics, threats and hazards.

It is important to note that there are four ways that regulatory change is effected:

Table 7-1. Applicability of Building Regulation to Terrorist Tactics, Threats and Hazards

Common Tactics	Zoning	Property Maintenance	Rehab	Construction
ATTACK DELIVERY				
Ballistic weapons				
Covert entry	●	●	●	●
Mail			●	●
Moving vehicle	●			
Stand-off weapons	●		●	●
Stationary vehicle	●	●	●	●
Supplies		●	●	●
ATTACK MECHANICS				
Airborne		●	●	●
Blast effects		●	●	●
Waterborne		●	●	●
THREATS/HAZARDS				
Armed attack		○	●	●
Arson/incendiary		●	●	●
Biological agent		●	●	●
Chemical agent		●	●	●
Conventional bomb	●	●	●	●
Cyber-terrorism			○	○
HAZMAT release	○	●	●	●
Nuclear device				○
Radiological agent		○	○	○
Surveillance				
Unauthorized entry				

LEGEND: ● = Applicability to designated type of regulation.   ○ = Possible applicability.

Source: FEMA

1.   Federal preemption
2.   State mandate or preemption
3.   Local prerogative
4.   Model code and voluntary standards

Any changes in the building regulatory system that would address the risk of terrorism will occur through these actions.

Developing codes and standards to deal with the risk of terrorism in new and existing buildings will require a broad acceptance of the risk, itself, and the effectiveness of the mitigation measures as well as a cost benefit assessment.

Sources
Department of Health and Human Services (DHHS), National Institute for Occupational Safety and Health (NIOSH), *Guidance for Protecting Building Environments from Air borne Chemical, Biological or Radiological Attacks*. DHHS (NIOSH) Publication No. 1 2002-139, May 2002.

Federal Emergency Management Agency (FEMA), *Insurance, Finance, and Regulation Primer for Terrorism Risk Management in Buildings*, FEMA 429, December, 2003.

Gustin, Joseph F. *Safety Management: A Guide for Facility Managers*, New York: UpWord Publishing, Inc., 1996.

Chapter 8

Assessing and Managing Risk

The terrorist attacks of September 11, 2001 clearly demonstrated that our institutions and our infrastructure have been targeted. The terrorist's weapons of mass destruction can include biological, chemical, radiological, nuclear, or high explosive weaponry. And, while terrorist groups would have to overcome significant technical and operational challenges to make and release many chemical and biological agents of a sufficient quality and quantity to kill large numbers of people, it has been tried.

Recent examples of terrorist activity, both at home and abroad, include the incidents of anthrax-laced letters sent to public and high-profile citizens following the September 11th attacks. In the release of the nerve agent sarin in a Tokyo subway in 1995, 12 people were killed and thousands were injured. In 1984, a religious cult in Oregon contaminated salad bars in local restaurants with salmonella bacteria in order to prevent people from voting in a local election. Although there were no fatalities as a result of this terrorist attack, hundreds of people were diagnosed with food-borne illness.

As both private and public sector organizations become more aware of their vulnerability to terrorist attacks, they are also becoming acutely aware of the need to increase security measures. In order to better prepare against terrorism and other threats, facility managers and building owners are reviewing their policies and procedures with an eye toward risk management.

RISK MANAGEMENT

Risk management is the systematic, analytical process that considers the likelihood of a threat harming individuals or physical assets. Risk management identifies actions that reduce the risk and mitigate the consequences of an attack or event. Risk management acknowledges that risk generally cannot be eliminated but risk can be reduced by enhancing protection from known or potential threats. These risk management principles:

• Identify weaknesses in a company, system or organization

• Offer a realistic method for making decisions about the expenditure of scarce resources and the selection of cost-effective countermeasures to protect assets;

• Improve the success rate of a company / organization's security efforts by emphasizing the communication of risks and recommendations to the final decision-making authority; and

• Assist facility managers, building owners and security professionals as well as other key decision makers answer the question "How much security is enough?"

Since risk is a function of assets, threats and vulnerabilities, risk management allows organizations and companies to determine the:

• magnitude and effect of potential loss;

• likelihood of such loss actually happening; and finally,

• countermeasures that could lower the probability or magnitude of the loss.

As defined by the National Infrastructure Protection Center (NIPC), a division of the Department of Homeland Security, risk is the potential for some unwanted event to occur. Examples of such unwanted events include loss of information and money, as well as organizational reputations, or someone gaining unauthorized access to the company's physical property, data systems, etc. As such, risk is the function of the likelihood of the unwanted event occurring and its consequences. Therefore, it is obvious that the higher the probability and the greater the consequences, the greater the risk to the company or organization.

The likelihood of the unwanted event occurring depends upon threat and vulnerability. Threat is the capability and the intention of a terrorist to undertake actions that are detrimental to a company or organization's interests. Threat is a function of the terrorist only. It cannot be controlled by the owner or the user of the asset, i.e., building owner, facility/property manager, etc. However, the intention of terrorists to exploit their capability may be encouraged by vulnerability in the company or organization's assets. Or conversely, the intention of the terrorist may be discouraged by the owner's countermeasures.

Vulnerability is any weakness in an asset that can be exploited by a terrorist to cause damage to the company/organization's interest. The level of vulnerability and level of risk can be reduced by implementing appropriate security countermeasures.

An asset is anything of value. Assets include:

• people
• information
• hardware
• software
• facilities
• company reputation
• company operations

In other words, assets are what a company or organization needs to get the job done. Consequently, the more critical the asset is to a company or organization, and the more critical the asset is to meeting company goals, the greater the effect of damage and/or destruction to the asset.

Risk Assessment

The first step in the process of assessing risk to a terrorist attack is to identify the relative importance of the people, business activities, goods and facilities involved in order to prioritize security actions. This applies to both new and existing facilities. The Federal Emergency Management Agency (FEMA) recommends the following actions:

• Define and understand the core functions and process of the business or institutional entity

• Identify critical business infrastructure:
— Critical component—people, functions and facilities
— Critical information systems and data

— Life safety systems and safe haven areas
— Security systems

- Assign a relative protection priority such as high, medium or low to the occupants, business functions or physical components of the facility:
  — High Priority—loss or damage of the facility would have grave consequences such as loss of life, severe injuries, loss of primary services or major loss of core processes and functions for an extended period of time.
  — Medium Priority—loss or damage of the facility would have moderate to serious consequences such as injuries or impairment of core functions and processes
  — Low Priority—loss or damage of the facility would have minor consequences or impact, such as a slight impact on core functions and processes for a short period of time.

THREAT ASSESSMENT

Terrorist threat comes from people with the intent to do harm, who are known to exist, have the capability for hostile action, and have expressed the intent to take hostile action.

Threat assessment is a continual process of compiling and examining information concerning potential threats. Information should be gathered from all reliable sources. The assessment process consists of:

- Defining threats and
- Identifying likely threat event profiles and tactics.

Defining Threats

Defining threats involves analyzing the following information regarding terrorists:

- Existence
- Capability
- History
- Intention
- Targeting

Existence is the assessment of who is hostile to the organization.

Capability is the assessment of what weapons have been used in carrying out past attacks.

History is the assessment of what the potential terrorist has done in the past and how many times.

Intention is the assessment of what the potential terrorist hopes to achieve.

Targeting is the assessment of the likelihood that a terrorist performing surveillance on a particular facility, nearby facilities, or facilities that have much in common with a particular organization.

Identifying Likely Threat Event Profiles and Tactics

To identify the likelihood of specific threats and tactics, the following variables should be evaluated:

- attack intentions
- hazard event profiles
- the expected effects of an attack on the facility/organization

Table 8-1 presents general event profiles for a range of possible forms of terrorist attacks. The profiles describe the mode, duration and extent of the effects of an attack, as well as any mitigating and exacerbating conditions that may exist. These descriptions can be used to identify threats of concern to individual organizations.

Assigning a Threat Rating

A threat rating should be assigned to each hazard of concern to a particular organization. The threat rating, like protection priority, is based on expert judgment. For purposes of simplicity the ratings may read high, medium or low:

- **High Threat**. Known terrorists or hazards, capable of causing loss and/or damage to a facility exist. One or more vulnerabilities are present and the terrorists are known or are reasonably suspected of having intent to attack the facility.

- **Medium Threat**. Known terrorists or hazards that may be capable of causing loss of or damage to a facility exists. One or more vulnerabilities may be present. However, the terrorists are not believed to have intent to attack the facility.

- **Low Threat**. Few or no terrorists or hazards exist. Their capability of causing damage to a particular facility is doubtful.

Table 8-1. Event Profiles for Terrorism and Technological Hazards

Section	Vulnerability Question	Guidance	Observations
1	Site		
1.1	What major structures surround the facility (site or building(s))? What critical infrastructure, government, military, or recreation facilities are in the local area that impact transportation, utilities, and collateral damage (attack at this facility impacting the other major structures or attack on the major structures impacting this facility)? What are the adjacent land uses immediately outside the perimeter of this facility (site or building(s))?	**Critical infrastructure to consider includes:** **Telecommunications infrastructure** Facilities for broadcast TV, cable TV; cellular networks; newspaper offices, production, and distribution; radio stations; satellite base stations; telephone trunking and switching stations, including critical cable routes and major rights-of-way **Electric power systems** Power plants, especially nuclear facilities; transmission and distribution system components; fuel distribution, delivery, and storage **Gas and oil facilities** Hazardous material facilities, oil/gas pipelines, and storage facilities	

Source: FEMA

These ratings may be changed over time. What may be a high threat in the present may, over time, be lessened to a medium or low threat depending upon the conditions at a particular time.

Alternative Approach

Assessing terrorist threats is the most difficult aspect of planning to resist terrorist attacks. This is particularly true for those building owners and facility managers who have not had any experience in doing so. An effective alternative approach may be to select a level of desired protection for a business operation based on management decision making, and then proceeding to a vulnerability assessment. Various federal agencies along with the U.S. Department of Defense correlate "levels of protection" with potential damage and expected injuries. The following levels are based on Department of Defense definitions:

- **High Protection**. Facility superficially damaged; no permanent deformation of primary and secondary structural members or non-structural elements. Only superficial injuries are likely.

- **Medium Protection**. Damaged, but repairable. Minor deformations of non-structural elements and secondary structural members and

Table 8-1. Event Profiles for Terrorism and Technological Hazards (*Cont'd*)

Hazard/Threat	Application Mode	Hazard Duration	Extent of Effects; Static/Dynamic	Mitigating and Exacerbating Conditions
Arson/Incendiary Attack	Initiation of fire or explosion on or near target via direct contact or remotely via projectile.	Generally minutes to hours.	Extent of damage is determined by type and quantity of device /accelerant and materials present at or near target. Effects generally static other than cascading consequences, incremental structural failure, etc.	Mitigation factors include built-in fire detection and protection systems and fire-resistive construction techniques. Inadequate security can allow easy access to target, easy concealment of an incendiary device and undetected initiation of a fire. Non-compliance with fire and building codes as well as failure to maintain existing fire protection systems can substantially increase the effectiveness of a fire weapon.
Biological Agents - Anthrax - Botulism - Brucellosis - Plague - Smallpox - Tularemia - Viral hemorrhagic fevers - Toxins (Botulinum, Ricin, Staphylococcal Enterotoxin B, T-2 Mycotoxins)	Liquid or solid contaminants can be dispersed using sprayers/aerosol generators or by point or line sources such as munitions, covert deposits, and moving sprayers.	Biological agents may pose viable threats for hours to years, depending on the agent and the conditions in which it exists.	Depending on the agent used and the effectiveness with which it is deployed, contamination can be spread via wind and water. Infection can be spread via human or animal vectors.	Altitude of release above ground can affect dispersion; sunlight is destructive to many bacteria and viruses; light to moderate winds will disperse agents but higher winds can break up aerosol clouds; the micro-meteorological effects of buildings and terrain can influence aerosolization and travel of agents.

Source: FEMA

no permanent deformation in primary structural members. Some minor injuries, but fatalities are unlikely.

- **Very Low Protection**. Heavily damaged, onset of structural collapse. Major deformation of primary and secondary structural members, but progressive collapse is unlikely. Collapse of non-structural elements. Majority of personnel suffer serious injuries. There are likely to be a limited number—10 percent to 25 percent—of fatalities.

Note that the "very low" level is not the same as doing nothing. No action could result in catastrophic building failure and high loss of life.

Table 8-1. Event Profiles for Terrorism and Technological Hazards (*Cont'd*)

Hazard/Threat	Application Mode	Hazard Duration	Extent of Effects; Static/Dynamic	Mitigating and Exacerbating Conditions
Chemical Agents - Blister - Blood - Choking/lung/pulmonary - Incapacitating - Nerve - Riot control/tear gas - Vomiting	Liquid/aerosol contaminants can be dispersed using sprayers or other aerosol generators; liquids vaporizing from puddles/ containers; or munitions.	Chemicals agents may pose viable threats for hours to weeks, depending on the agent and the conditions in which it exists.	Contamination can be carried out of the initial target area by persons, vehicles, water, and wind. Chemicals may be corrosive or otherwise damaging over time if not remediated.	Air temperature can affect evaporation of aerosols. Ground temperature affects evaporation of liquids. Humidity can enlarge aerosol particles, reducing inhalation hazard. Precipitation can dilute and disperse agents, but can spread contamination. Wind can disperse vapors, but also cause target are to be dynamic. The micro-meteorological effects of buildings and terrain can alter travel and duration of agents. Shielding in the form of sheltering in place can protect people and property from harmful effects.
Conventional Bomb - Stationary vehicle - Moving vehicle - Mail - Supply - Thrown - Placed - Personnel	Detonation of explosive device on or near target; via person, vehicle, or projectile.	Instantaneous; additional secondary devices may be used, lengthening the time duration of the hazard until the attack site is determined to be clear.	Extent of damage is determined by type and quantity of explosive. Effects generally static other than cascading consequences, incremental structural failure, etc.	Energy decreases logarithmically as a function of distance from seat of blast. Terrain, forestation, structures, etc., can provide shielding by absorbing and/or deflecting energy and debris. Exacerbating conditions include ease of access to target; lack of barriers/shielding; poor construction; and ease of concealment of device.
Cyberterrorism	Electronic attack using one computer system against another.	Minutes to days.	Generally no direct effects on built environment.	Inadequate security can facilitate access to critical computer systems, allowing them to be used to conduct attacks.

Source: FEMA

Table 8-1. Event Profiles for Terrorism and Technological Hazards (*Cont'd*)

Hazard/Threat	Application Mode	Hazard Duration	Extent of Effects; Static/Dynamic	Mitigating and Exacerbating Conditions
Hazardous Material Release (fixed facility or transportation) - Toxic Industrial Chemicals and Materials (Organic vapors: cyclohexane; Acid gases: cyanogens, chlorine, hydrogen sulfide; Base gases: ammonia; Special cases: phosgene, formaldehyde)	Solid, liquid, and/or gaseous contaminants may be released from fixed or mobile containers.	Hours to days.	Chemicals may be corrosive or otherwise damaging over time. Explosion and/or fire may be subsequent. Contamination may be carried out of the incident area by persons, vehicles, water, and wind.	As with chemical weapons, weather conditions will directly affect how the hazard develops. The micro-meteorological effects of buildings and terrain can alter travel and duration of agents. Shielding in the form of sheltering in place can protect people and property from harmful effects. Non-compliance with fire and building codes as well as failure to maintain existing fire protection and containment features can substantially increase the damage from a hazardous materials release.
Nuclear Device	Detonation of nuclear device underground, at the surface, in the air or at high altitude.	Light/heat flash and blast/shock wave last for seconds; nuclear radiation and fallout hazards can persist for years. Electromagnetic pulse from a high-altitude detonation lasts for seconds and affects only unprotected electronic systems.	Initial light, heat and blast effects of a subsurface, ground or air burst are static and are determined by the device's characteristics and employment; fallout of radioactive contaminants may be dynamic, depending on meteorological conditions.	Harmful effects of radiation can be reduced by minimizing the time of exposure. Light, heat, and blast energy decrease logarithmically as a function of distance from seat of blast. Terrain, forestation, structures, etc., can provide shielding by absorbing and/or deflecting radiation and radioactive contaminants.
Radiological Agents - Alpha - Beta - Gamma	Radioactive contaminants can be dispersed using sprayers/aerosol generators, or by point or line sources such as munitions, covert deposits, and moving sprayers.	Contaminants may remain hazardous for seconds to years, depending on material used.	Initial effects will be localized to site of attack; depending on meteorological conditions, subsequent behavior of radioactive contaminants may be dynamic.	Duration of exposure, distance from source of radiation, and the amount of shielding between source and target determine exposure to radiation.

Source: FEMA

Table 8-1. Event Profiles for Terrorism and Technological Hazards (*Cont'd*)

Hazard/Threat	Application Mode	Hazard Duration	Extent of Effects; Static/Dynamic	Mitigating and Exacerbating Conditions
Surveillance - Acoustic - Electronic eavesdropping - Visual	Stand-off collection of visual information using cameras or high powered optics, acoustic information using directional microphones and lasers, and electronic information from computers, cell phones, and hand-held radios. Placed collection by putting a device "bug" at the point of use.	Usually months.	This is usually the prelude to the loss of an asset. A terrorist surveillance team spends much time looking for vulnerabilities and tactics that will be successful. This is the time period that provides the best assessment of threat as it indicates targeting of the facility.	Building design, especially blocking lines of sight and ensuring the exterior walls and windows do not allow sound transmission or acoustic collection, can mitigate this hazard.
Unauthorized Entry - Forced - Covert	Use of hand or power tools, weapons, or explosives to create a man-sized opening or operate an assembly (such as a locked door), or use false credentials to enter a building.	Minutes to hours, depending upon the intent.	If goal is to steal or destroy physical assets or compromise information, the initial effects are quick, but damage may be long lasting. If intent is to disrupt operations or take hostages, the effects may last for a long time, especially if injury or death occurs.	Standard physical security building design should be the minimum mitigation measures. For more critical assets, additional measures, like closed circuit television or traffic flow that channels visitors past access control, aids in detection of this hazard.

Source: FEMA

VULNERABILITY ASSESSMENT

A terrorism vulnerability assessment evaluates any weaknesses that can be exploited by a terrorist. It evaluates the vulnerability of facilities across a broad range of identified threats/hazards and provides a basis for determining physical and operational mitigation measures for their protection. It applies both to new building programming and design as well as to existing building management and renovation over the service life of a structure.

A vulnerability rating can also be assigned to the appropriate aspects of building operations and systems to the defined threats for the particular facility. These ratings can also be assigned as high, medium or low:

- **High Vulnerability.** One or more significant weaknesses have been identified that make the facility highly susceptible to a terrorist or hazard.

- **Medium Vulnerability.** A weakness has been identified that makes the facility somewhat susceptible to a terrorist or hazard.

- **Low Vulnerability.** A minor weakness has been identified that slightly increases the susceptibility of the facility to a terrorist or hazard.

The Building Vulnerability Assessment Checklist located in the Appendix compiles a list of questions to be addressed in assessing the vulnerability of facilities to terrorist attack. The following is useful in the initial screening of existing facilities to identify and prioritize terrorism risk reduction needs.

Initial Vulnerability Estimate

The initial vulnerability estimate provides a quick, qualitative assessment of the vulnerability of existing buildings to terrorist attack. Three means of data collection using a simple scale of high, medium and low ratings may provide useful information. The data collection is based upon three criteria:

- Visual inspection
- Document review
- Organization and management procedures review

Visual Inspection—When visually inspecting the condition of the property, an evaluation of the site and all the facility systems is performed. This includes the architectural, structural, building envelope, utility, mechanical, plumbing and gas, electrical, fire alarm, communications and information technology systems. Equipment operations and maintenance procedures and records, security systems, and planning and procedures should also be evaluated. The inspection may need to go beyond the site to determine the vulnerability of utility and other infrastructure systems.

Document Review—The planning team, which includes the building owner and the facility manager, should review all necessary plans, specifications and related construction data in terms of terrorism vulnerability. Equipment operation, maintenance procedures and records, as well as security procedures, should be included in this review.

Organization and Management Procedures Review—The planning team should review business and operations practices and procedures to identify opportunities that can reduce exposure to attack. This review also includes tenant operations.

Vulnerability Estimate Screening

The following screening table from FEMA provides guidance for initial vulnerability assessment. The goal of this assessment is to distinguish facilities of high, medium or low vulnerability to terrorist attack. The implication is that high vulnerability facilities should receive more detailed analysis. Specific strategies for risk reduction should be developed.

For this initial assessment, subjective ratings by building owners, facility managers and other qualified professionals who are familiar with the facility, are appropriate. Assigning a "high", "medium" or "low" vulnerability rating to the responses to vulnerability questions for each building system, will provide a preliminary basis for estimating the overall vulnerability of a particular facility to terrorist attack. The responses will also indicate areas of opportunity for mitigation actions to reduce terrorism risk.

Site Questions

Vulnerability assessment of the "site" examines surrounding structures, terrain, perimeter controls, traffic patterns and separations, landscaping elements/features, and lines of sight.

"Site" questions focus primarily on visual inspection to develop ratings. The questions emphasize vulnerability to moving vehicle, stationary vehicle, and covert entry tactics. Vulnerability to blast is the primary concern addressed.

Architectural Questions

Assessing "architectural" vulnerability investigates tenancy, services, public and private access, access controls, activity patterns and exposures.

"Architectural" questions focus equally on visual inspection and

Table 8-2. Site Systems Vulnerability Estimate

Question	Vulnerability Rating (H, M, L)	Visual inspection	Document review	Org/Mgmt procedure	Moving vehicle	Stationary vehicle	Covert entry	Mail	Supplies	Blast effects	Airborne (contamination)	Waterborne (contamination)
What major structures surround the facility?	□	•	•									
What critical infrastructure, government, military, or recreation facilities are in the local area that impact transportation, utilities, and collateral damage (attack at this facility impacting the other major structures or attack on the major structures impacting this facility)?	□	•	•	•								
What are the adjacent land uses immediately outside the perimeter of this facility?	□	•	•									
What are the site access points to the facility?	□	•			•		•					
What is the minimum distance from the inspection location to the building?	□	•			•		•			•		
Is there any potential access to the site or facility through utility paths or water runoff?	□	•	•				•					
What are the existing types of vehicle anti-ram devices for the facility?	□	•			•					•		
What is the anti-ram buffer zone standoff distance from the building to unscreened vehicles or parking?	□	•			•							
Are perimeter barriers capable of stopping vehicles?	□	•	•		•							
Does site circulation prevent high-speed approaches by vehicles?	□	•			•							
Is there a minimum setback distance between the building and parked vehicles?	□	•				•				•		
Does adjacent surface parking maintain a minimum standoff distance?	□	•				•				•		
Do site landscaping and street furniture provide hiding places?	□	•					•					

LEGEND: □ = Determine high, medium, or low vulnerability rating. • = Applicability of factor to question.

Source: FEMA

evaluation of organizational and management procedures to develop ratings. The questions emphasize vulnerability to moving vehicle, stationary vehicle and covert entry tactics. Vulnerability to blast is a primary concern.

Structural and Building Envelope Questions

A vulnerability assessment of "structural" systems examines construction type, materials, detailing, collapse characteristics and critical elements. An assessment of the "building envelope" examines strength, fenestration, glazing characteristics/detailing and anchorage.

Table 8-3. Architectural Systems Vulnerability Estimate

	Vulnerability Rating (H, M, L)	Visual inspection	Document review	Org/Mgmt procedure	Moving vehicle	Stationary vehicle	Covert entry	Mail	Supplies	Blast effects	Airborne (contamination)	Waterborne (contamination)
What major structures surround the facility?	□	•	•	•								
Do entrances avoid significant queuing?	□	•		•								
What are the adjacent land uses immediately outside the perimeter of this facility?	□	•		•								
Are public and private activities separated?	□	•					•					
Are critical assets (people, activities, building systems and components) located close to any main entrance, vehicle circulation, parking, maintenance area, loading dock, or interior parking?	□	•	•	•	•	•	•	•			•	
Are high-value or critical assets located as far into the interior of the building as possible and separated from the public areas of the building?	□	•	•	•							•	
Is high visitor activity away from critical assets?	□			•			•					
Are critical assets located in spaces that are occupied 24 hours per day?	□						•					
Are assets located in areas where they are visible to more than one person?	□						•					
Do interior barriers differentiate level of security within a facility?	□	•	•	•								
Are emergency systems located away from high-risk areas?	□	•	•	•								

LEGEND: □ = Determine high, medium, or low vulnerability rating.   • = Applicability of factor to question.

Source: FEMA

These questions rely on a review of construction documents and visual inspection to develop ratings. Vulnerability to blasts is the primary concern.

Utility Systems Questions

A vulnerability assessment of "utility" systems examines the full range of source and supply systems serving the facility including water, fuel and the electricity supply, as well as the fire alarm and suppression systems and communication systems.

These questions rely on information obtained from visual inspection, review of construction documents and organizational/management procedures to develop the ratings. Vulnerability to waterborne contaminants is the primary consideration.

Mechanical Systems

A vulnerability assessment of mechanical systems examines air supply and exhaust configurations, filtration, sensing and monitoring, system zoning and control, and elevator management.

These questions and ratings rely on information obtained from review of construction documents and visual inspection. Vulnerability to airborne contaminants is the primary concern, including contamination from chemical, biological and radiological attack.

Plumbing and Gas Systems

A vulnerability assessment of plumbing and gas systems examines the liquid distribution systems serving the facility including water and fuel distribution, water heating and fuel storage.

These questions rely on information from a review of construction documents to develop ratings. Vulnerability to waterborne contaminants is the primary concern.

Electrical Systems

A vulnerability assessment of electrical systems examines transformer and switchgear security, electricity distribution and accessibility, as well

Table 8-4. Structural and Building Envelope Systems Vulnerability Estimate

FEMA 'Structural & Building Envelope Systems' Vulnerability Estimate	Vulnerability Rating (H, M, L)	Visual inspection	Document review	Org/Mgmt procedure	Moving vehicle	Stationary vehicle	Covert entry	Mail	Supplies	Blast effects	Airborne (contamination)	Waterborne (contamination)
What type of construction?	❑	●	●								●	
Is the column spacing minimized so that reasonably sized members will resist the design loads and increase the redundancy of the system?	❑	●	●								●	
What are the floor-to-floor heights?	❑	●	●								●	
Is the structure vulnerable to progressive collapse?	❑	●	●								●	
Are there adequate redundant load paths in the structure?	❑	●	●								●	
What is the designed or estimated protection level of the exterior walls against the postulated explosive threat?	❑		●								●	

LEGEND: ❑ = Determine high, medium, or low vulnerability rating.   ● = Applicability of factor to question.

Source: FEMA

Table 8-5. Utility Systems Vulnerability Estimate

	Vulnerability Rating (H, M, L)	Visual inspection	Document review	Org/Mgmt procedure	Moving vehicle	Stationary vehicle	Covert entry	Mail	Supplies	Blast effects	Airborne (contamination)	Waterborne (contamination)
What is the source of domestic water? (utility, municipal, wells, lake, river, storage tank)	❑	●	●									●
How many gallons and how long will it allow operations to continue?	❑	●	●	●								●
What is the source of water for the fire suppression system? (local utility company lines, storage tanks with utility company backup, lake, or river)	❑	●	●									
Are there alternate water supplies for fire suppression?	❑	●	●	●								
Are the sprinkler and standpipe connections adequate and redundant?	❑	●	●									
What fuel supplies do the facility rely upon for critical operation?	❑	●	●	●								
Where is the fuel supply obtained?	❑		●									
Are there alternate sources of fuel?	❑		●									
Can alternate fuels be used?	❑		●	●								
What is the normal source of electrical service for the facility?	❑	●	●									
What provisions for emergency power exist? What systems receive emergency power and have capacity requirements been tested?	❑	●	●	●								
By what means does the main telephone and data communications interface the facility?	❑	●	●	●								

LEGEND: ❑ = Determine high, medium, or low vulnerability rating.   ● = Applicability of factor to question.

Source: FEMA

as emergency systems.

These questions rely on information from visual inspection and a review of construction documents to develop ratings. No particular attack mechanism is emphasized.

Fire Alarm Systems

A vulnerability assessment of fire alarm systems examines detection sensing and signaling, system configurations, accessibility of controls and redundancies.

These questions rely on information obtained from the review of the construction documents, as well as a review of organizational/ management procedures to develop ratings. No particular attack mechanism is emphasized.

Table 8-6. Mechanical Systems Vulnerability Estimate

	Vulnerability Rating (H, M, L)	Visual inspection	Document review	Org/Mgmt procedure	Moving vehicle	Stationary vehicle	Covert entry	Mail	Supplies	Blast effects	Airborne (contamination)	Waterborne (contamination)
Where are the air intakes and exhaust louvers for the building? (low, high, or midpoint of the building structure)	□	●	●								●	
Are there multiple air intake locations?	□	●	●								●	
How are air handling systems zoned?	□	●	●								●	
Are there large central air handling units or are there multiple units serving separate zones?	□		●								●	
Are there any redundancies in the air handling system?	□			●	●						●	
Where is roof-mounted equipment located on the roof? (near perimeter, at center of roof)	□	●										

LEGEND: □ = Determine high, medium, or low vulnerability rating.   ● = Applicability of factor to question.

Source: FEMA

Table 8-7 Plumbing and Gas Systems Vulnerability Estimate

	Vulnerability Rating (H, M, L)	Visual inspection	Document review	Org/Mgmt procedure	Moving vehicle	Stationary vehicle	Covert entry	Mail	Supplies	Blast effects	Airborne (contamination)	Waterborne (contamination)
What is the method of water distribution?	□	●										●
What is the method of gas distribution? (heating, cooking, medical, process)	□	●										
What is the method of heating domestic water?	□	●	●	●								
Are there reserve supplies of critical gases?	□		●	●								

LEGEND: □ = Determine high, medium, or low vulnerability rating.   ● = Applicability of factor to question.

Source: FEMA

Communications and Information Technology Systems

A vulnerability assessment of communications and information technology systems examines distribution, power supplies, accessibility, control, notification and backups.

These questions rely on information from visual inspection, a review

Table 8-8. Electrical Systems Vulnerability Estimate

	Vulnerability Rating (H, M, L)	Visual inspection	Document review	Org/Mgmt procedure	Moving vehicle	Stationary vehicle	Covert entry	Mail	Supplies	Blast effects	Airborne (contamination)	Waterborne (contamination)
Are there any transformers or switchgears located outside the building or accessible from the building exterior?	❑	●										
Are they (transformers or switchgears) vulnerable to public access?	❑	●										
Are critical electrical systems located in areas outside of secured electrical areas?	❑	●	●	●								
Does emergency backup power exist for all areas within the facility or for critical areas only?	❑	●	●									

LEGEND: ❑ = Determine high, medium, or low vulnerability rating. ● = Applicability of factor to question.

Source: FEMA

Table 8-9. Fire Alarm Systems Vulnerability Estimate

	Vulnerability Rating (H, M, L)	Visual inspection	Document review	Org/Mgmt procedure	Moving vehicle	Stationary vehicle	Covert entry	Mail	Supplies	Blast effects	Airborne (contamination)	Waterborne (contamination)
Is the fire alarm system stand-alone or integrated with other functions such as security and environmental or building management systems?	❑		●	●								
Is there redundant off-premises fire alarm reporting?	❑		●	●								

LEGEND: ❑ = Determine high, medium, or low vulnerability rating. ● = Applicability of factor to question.

Source: FEMA

of construction documents, as well as a review of the organizational/ management procedures to develop ratings. No particular attack mechanism is emphasized.

As companies and organizations increase their security measures and attempt to identify vulnerabilities in critical assets, they are looking for a mechanism to ensure an efficient investment of resources to counter threats. One such mechanism is a risk management model that will:

Table 8-10. Communication and its Systems Vulnerability Estimate

	Vulnerability Rating (H, M, L)	Visual inspection	Document review	Org/Mgmt procedure	Moving vehicle	Stationary vehicle	Covert entry	Mail	Supplies	Blast effects	Airborne (contamination)	Waterborne (contamination)
Where is the main telephone distribution room and where is it in relation to higher risk areas?	▢	●	●	●								
Where are communication systems wiring closets located? (voice, data, signal, alarm)	▢	●	●									

LEGEND: ▢ = Determine high, medium, or low vulnerability rating. ● = Applicability of factor to question.

Source: FEMA

- assess assets, threats and vulnerabilities; and
- incorporate a continuous assessment feature.

The questions listed above provide a framework for such a model. By reviewing these ratings, a preliminary determination can be made of where the major vulnerabilities and threats could occur in a facility. Additionally, a preliminary determination can be made regarding which risks require immediate attention. As noted, however, the ratings (low, medium, high) are subjective and as a result may not be exact. A more detailed and quantitative evaluation is required. This detailed evaluation involves a significantly more thorough review of information in all areas, including additional information that concerns:

- Equipment Operations and Maintenance (up-to-date drawings, manuals and procedures, training, monitoring, etc.);

- Security Systems (perimeter and interior sensing, monitoring and control, security system documentation and training, etc.); and

- The Security Master Plan (currency, responsibilities, etc.).

A complete list of detailed questions can be found in Appendix II. These questions should be considered in fully evaluating vulnerability to terrorist threats. The means of data collection that should be employed and the particular tactics and attack mechanisms addressed by each question are identified in the appendix. This is done so that specialized checklists can be created to assess vulnerability to terrorist tactics of particular concern to an individual company or organization.

By following such a model, organizations are able to "tailor" their management of risk to the current situation, as well as to assess future risks.

Sources

Federal Emergency Management Agency (FEMA), *Risk Management Series: Insurance, Finance, and Regulation Primer for Terrorism Risk Management in Buildings*, FEMA 429. December 2003.

Federal Emergency Management Agency (FEMA), *Risk Management Series: Primer for Design of Commercial Buildings to Mitigate Terrorist Attacks*, FEMA 427. December 2003.

Gustin, Joseph, *Cyberterrorism: A Guide for Facility Managers*, Lilburn, GA: The Fairmont Press, Inc., 2004.

Gustin, Joseph F. *Disaster and Recovery Planning: A Guide for Facility Managers*, 3rd ed., Lilburn, GA: The Fairmont Press, Inc., 2004.

National Infrastructure Protection Center (NIPC). *Risk Management: An Essential Guide to Protecting Critical Assets*, November 2002.

Chapter 9

Crisis Management

E vents over the past several years have clearly demonstrated the need to incorporate new threats into emergency management planning. The September 11, 2001 attacks on New York City, Washington, D.C., the Fall, 2001 anthrax attacks, the 1995 bombing of the Murrah Federal Building in Oklahoma City and the 1993 World Trade Center bombing are prime examples of the need to reduce vulnerability to terrorist acts.

As previously noted "terrorism" refers to intentional, criminal and malicious acts. While there is no single universally accepted definition of terrorism, it can be interpreted in many ways. Officially, terrorism is defined in the Code of Federal Regulations as "…the unlawful use of force and violence against persons or property to intimidate or coerce a government, the civilian population, or any segment thereof, in furtherance of political or social objectives." (28 CFR, Section 0.85). The Federal Bureau of Investigation (FBI) further characterizes terrorism as either domestic or international depending upon the origin, base and objectives of the terrorist organization. However, the origin of the terrorist or person causing the hazard is far less relevant to mitigation planning than the hazard itself and its consequences.

As such, terrorism refers to the use of weapons of mass destruction (WMD). This includes biological, chemical, nuclear and radiological weapons; arson, incendiary, explosive, and armed attacks; industrial sabotage and intentional hazardous materials releases; and cyberterrorism. There are variations within these categories. For example, in the area of biological and chemical weapons, there are numerous agents and ways for these agents to be disseminated.

The events listed above clearly suggest that mitigation planning should also address hazards generated by human activities such as terrorism. The term mitigation refers generally to activities that reduce the loss of life and property by either reducing or eliminating the effects of disasters. In the context of terrorism, mitigation is in the often interpreted to include a wide variety of preparedness and response actions.

Mitigation in its traditional sense refers to those specific actions that can be taken to reduce loss of life and property by reducing risk. Those specific actions are best described in the phases of emergency management:

* Mitigation
* Preparedness
* Response
* Recovery

Mitigation is defined as any sustained action that is taken to reduce or eliminate long-term risk to life and property from a hazard event. It is also known as prevention when done before a disaster. As such, it encourages long-term reduction of hazard vulnerability. The goal of mitigation is to decrease the need for response as opposed to increasing the response capability. Mitigation can save lives and reduce property damage, and should be cost-effective and environmentally sound. This in turn can reduce the cost of disasters to property owners as well as to all levels of government. In addition, mitigation can protect facilities, reduce exposure to liability and minimize disruption.

Preparedness includes plans and preparation made to save lives and property as well as to facilitate response operations.

Response includes those actions taken to provide emergency assistance, save lives, minimize property damage and speed recovery immediately following an occurrence.

Recovery includes actions taken to return to normal operations following an occurrence.

As with any other form of disaster, a deliberate terrorist attack may not always be preventable. However, effective mitigation planning can reduce the likelihood and/or the potential effects of the occurrence. The mitigation planning process involves the following:

* Identifying and organizing resources—the planning team

* Conducting a risk or threat assessment and estimating potential losses

* Identifying mitigation actions that will reduce the effects of the hazard

* Implementing the actions, evaluating results and updating the plan

Planning, in its larger sense, helps to coordinate actions, accomplish goals, leverage opportunities and identify priorities for resource allocations. Hazard mitigation planning integrates these activities into emergency management programming.

HAZARD MITIGATION PLANNING: THE PROCESS

Identifying and Organizing Resources—
Building the Planning Team

Identifying and organizing resources addresses the creation of a planning team. There must be an individual or group in charge of developing the plan. The following provides guidance for forming the planning team.

Forming the Team

The size of the planning team will depend on the facility's operation, requirements and resources. Involving a group of people is particularly effective because it:

* Encourages participation and gets more people invested in the process;
* Increases the amount of time and energy participants are able to give;
* Enhances the visibility and stature of the planning process; and
* Provides for a broad perspective on the issues.

While some persons will serve as active members and others will serve in advisory capacities, input from all functional areas within the company should be obtained. These other areas include:

* Upper management
* Line management
* Labor
* Human resources
* Engineering and maintenance
* Safety, health and environmental affairs
* Public relations/information
* Security

- Community relations
- Sales and marketing
- Legal
- Finance and purchasing

Regardless of the role they serve, all planning team members should be appointed in writing by upper management. Their job descriptions could also reflect their assignment.

External resources that will be helpful in addressing human engineered attacks or incidents include:

- Chemical emergency planning
- Law enforcement and crime prevention
- Electrical engineering
- Emergency management
- Explosives/blast characteristics
- Fire protection engineering
- Internal security
- Mechanical engineering including HVAC
- Site planning/urban design/landscape design
- Structural engineering, design and construction

Specialized expertise in these fields can be found in a number of community resources. Additionally, technical assistance from the federal government may be available. Among the federal organizations offering relevant support are the Department of Homeland Security (DHS), the Environmental Protection Agency (EPA), and the Department of Justice (DOJ). Appendix V contains web links to various government agency programs.

Risk Assessment

Assessing risks involves identifying hazards and estimating potential losses. There are some unique aspects to hazard characteristics, asset identification, and vulnerability assessment that will affect the way a risk assessment for terrorism is carried out.

Human engineered hazards fall into two general categories: 1). terrorism, or intentional acts, and 2). technological hazards, or accidental events. In terms of terrorism, the following hazards are included:

- Conventional bombs/improvised explosive devices
- Biological agents
- Chemical agents
- Nuclear bombs
- Radiological agents
- Arson/incendiary attacks
- Armed attacks
- Cyberterrorism
- Agriterrorism
- Intentional hazardous materials release

In terms of technological or accidental occurrences, the following hazards are included:

- Industrial accidents (facility)
- Industrial accidents (transportation)
- Failure of Supervisory Control and Data Acquisition (SCADA) system, or other critical infrastructure components.

Within these incidents, or occurrences, there are many variations. These variations illustrate one of the basic differences between natural and human-engineered hazards. For example, the types, frequencies and locations of many natural hazards are identifiable. In some cases, they are predictable. However, carelessness, incompetence and other behaviors are functions of the human mind and cannot be predicted with any degree of accuracy.

Hazard Event Profiling

There are significant differences been natural and human engineered hazards, particularly those related to terrorism. The primary difference lies with the fact that terrorists have the ability to choose their targets and their tactics. In doing so, they can design their attack to maximize the changes of achieving their objectives. This makes it difficult to identify how and where the risk occurs. Notwithstanding this difficulty, the consequences of these intentional occurrences are generally familiar to the emergency planning and response community. These authoritative community sources can provide detailed information on injuries and deaths, building contamination and/or damage to buildings and systems, etc. More importantly, they can also provide details about the various agents' characteristics and the ways in which these hazards can impact facilities, as well as the actions that can be taken to reduce or eliminate the resulting damage.

Human-engineered occurrences involve one or more methods of harmful force to facilities. These methods include biological, chemical, radiological or nuclear hazard contamination; energy release from explosives, arson and electromagnetic waves; or failure or denial of service through sabotage, infrastructure breakdown and transportation service disruption. As such, the planning team should include expertise in these areas in order to develop a comprehensive list of potential human-engineered hazards, as well as to identify the way(s) in which the hazards might occur.

The Federal Emergency Management Agency (FEMA) provides information to help guide and provide a supplement to building the planning team. Table 9-1 suggest additional individuals, agencies and organizations that should be included on a team to plan for manmade hazards. State organizations can be included on local teams when appropriate to serve as a source of information and to provide guidance and coordination. Table 9-1 could be used as a starting point for expanding the team.

This support can assist the team in understanding how human-engineered/terrorist activities can affect a facility. It is not intended to take the place of the advice and recommendations given by security, planning or design professionals. Note that for each type of hazard, FEMA identifies a set of factors that describes how facilities are affected by a particular hazard. The definitions for these factors are:

• *Application* mode describes the human act(s) or unintended event(s) necessary to cause the hazard to occur.

• *Duration* is the length of time the hazard is present on the target. For example, the duration of a tornado may be just minutes, but a chemical warfare agent such as mustard gas, if not remediated, can persist for days or weeks under the right conditions.

• The *dynamic/static characteristic* of a hazard describes its tendency, or that of its effects, to either expand, contract, or remain confined in time, magnitude and space. For example, the physical destruction caused by an earthquake is generally confined to the place in

Table 9-1. Build The Planning Team

	ON TEAM	ADD TO TEAM		ON TEAM	ADD TO TEAM
Specialists for Manmade Hazards			**Special Districts and Authorities**		
Bomb and Arson Squads	☐	☐	Airport and Seaport Authorities	☐	☐
Community Emergency Response Teams	☐	☐	Business Improvement District(s)	☐	☐
Hazardous Materials Experts	☐	☐	Fire Control District	☐	☐
Infrastructure Owners/Operators	☐	☐	Flood Control District	☐	☐
National Guard Units	☐	☐	Redevelopment Agencies	☐	☐
Representatives from facilities identified in Worksheet #2: Asset Identification Checklist	☐	☐	Regional/Metropolitan Planning Organization(s)	☐	☐
			School Districts	☐	☐
Local/Tribal			Transit/Transportation Agencies	☐	☐
Administrator/Manager's Office	☐	☐	**Others**		
Budget/Finance Office	☐	☐	Architectural/Engineering/Planning Firms	☐	☐
Building Code Enforcement Office	☐	☐	Citizen Corps	☐	☐
City/County Attorney's Office	☐	☐	Colleges/Universities	☐	☐
Economic Development Office	☐	☐	Land Developers	☐	☐
Emergency Preparedness Office	☐	☐	Major Employers/Businesses	☐	☐
Fire and Rescue Department	☐	☐	Professional Associations	☐	☐
Hospital Management	☐	☐	Retired Professionals	☐	☐
Local Emergency Planning Committee	☐	☐			
Planning and Zoning Office	☐	☐	**State**		
Police/Sheriff's Department	☐	☐	Adjutant General's Office (National Guard)	☐	☐
Public Works Department	☐	☐	Board of Education	☐	☐
Sanitation Department	☐	☐	Building Code Office	☐	☐
School Board	☐	☐	Climatologist	☐	☐
Transportation Department	☐	☐	Earthquake Program Manager	☐	☐
Tribal Leaders	☐	☐	Economic Development Office	☐	☐

Source: FEMA

(Continued)

which it occurs, and it does not usually get worse unless there are aftershocks or other cascading failures. In contrast, a cloud of chlorine gas leaking from a storage tank can change location by drifting with the wind and can diminish in danger by dissipating over time.

- *Mitigating conditions* are characteristics of the target and its physical environment that can reduce the effects of a hazard. For example, earthen berms can provide protection from bombs; exposure to sunlight can render some biological agents ineffective;

Table 9-1. Build A Planning Team (*Cont'd*)

	ON TEAM	ADD TO TEAM		ON TEAM	ADD TO TEAM
Emergency Management Office/ State Hazard Mitigation Officer	☐	☐	**Non-Governmental Organizations (NGOs)**		
Environmental Protection Office	☐	☐	American Red Cross	☐	☐
Fire Marshal's Office	☐	☐	Chamber of Commerce	☐	☐
Geologist	☐	☐	Community/Faith-Based Organizations	☐	☐
Homeland Security Coordinator's Office	☐	☐	Environmental Organizations	☐	☐
Housing Office	☐	☐	Homeowners Associations	☐	☐
Hurricane Program Manager	☐	☐	Neighborhood Organizations	☐	☐
Insurance Commissioner's Office	☐	☐	Private Development Agencies	☐	☐
National Flood Insurance Program Coordinator	☐	☐	Utility Companies	☐	☐
Natural Resources Office	☐	☐	Other Appropriate NGOs	☐	☐
Planning Agencies	☐	☐			
Police	☐	☐			
Public Health Office	☐	☐			
Public Information Office	☐	☐			
Tourism Department	☐	☐			

and effective perimeter lighting and surveillance can minimize the likelihood of someone approaching a target unseen. In contrast, exacerbating conditions are characteristics that can enhance or magnify the effects of a hazard. For example, depressions or low areas in terrain can trap heavy vapors, and a proliferation of street furniture—trash receptacles, newspaper vending machines, mail boxes, etc.—can provide concealment opportunities for explosive devices.

Inventory Assets

As noted earlier, the probability of a human-engineered/terrorist occurrence cannot be "forecast" with any significant level of accuracy. Subsequently, the planning team should use an asset-specific approach. This approach can identify potentially at-risk critical assets and systems.

Once a comprehensive list of assets has been developed, it should be prioritized in order to ensure that the most critical assets are protected first. Following the prioritization of the most critical assets, the vulnerabilities of each facility or system to each type of hazard should be assessed.

Table 9-2. Event Profiles for Terrorism

Hazard	Application Mode	Hazard Duration	Extent of Effects; Static/Dynamic	Mitigating and Exacerbating Conditions
Conventional Bomb/ Improvised Explosive Device	Detonation of explosive device on or near target; delivery via person, vehicle, or projectile.	Instantaneous; additional "secondary devices" may be used, lengthening the time duration of the hazard until the attack site is determined to be clear.	Extent of damage is determined by type and quantity of explosive. Effects generally static other than cascading consequences, incremental structural failure, etc.	Overpressure at a given standoff is inversely proportional to the cube of the distance from the blast; thus, each additional increment of standoff provides progressively more protection. Terrain, forestation, structures, etc. can provide shielding by absorbing and/or deflecting energy and debris. Exacerbating conditions include ease of access to target; lack of barriers/shielding; poor construction; and ease of concealment of device.
Chemical Agent	Liquid/aerosol contaminants can be dispersed using sprayers or other aerosol generators; liquids vaporizing from puddles/ containers; or munitions.	Chemical agents may pose viable threats for hours to weeks depending on the agent and the conditions in which it exists.	Contamination can be carried out of the initial target area by persons, vehicles, water and wind. Chemicals may be corrosive or otherwise damaging over time if not remediated.	Air temperature can affect evaporation of aerosols. Ground temperature affects evaporation of liquids. Humidity can enlarge aerosol particles, reducing inhalation hazard. Precipitation can dilute and disperse agents but can spread contamination. Wind can disperse vapors but also cause target area to be dynamic. The micro-meteorological effects of buildings and terrain can alter travel and duration of agents. Shielding in the form of sheltering in place can protect people and property from harmful effects.
Arson/ Incendiary Attack	Initiation of fire or explosion on or near target via direct contact or remotely via projectile.	Generally minutes to hours.	Extent of damage is determined by type and quantity of device/accelerant and materials present at or near target. Effects generally static other than cascading consequences, incremental structural failure, etc.	Mitigation factors include built-in fire detection and protection systems and fire-resistive construction techniques. Inadequate security can allow easy access to target, easy concealment of an incendiary device and undetected initiation of a fire. Non-compliance with fire and building codes as well as failure to maintain existing fire protection systems can substantially increase the effectiveness of a fire weapon.
Armed Attack	Tactical assault or sniping from remote location.	Generally minutes to days.	Varies based upon the perpetrators' intent and capabilities.	Inadequate security can allow easy access to target, easy concealment of weapons and undetected initiation of an attack.
Biological Agent	Liquid or solid contaminants can be dispersed using sprayers/aerosol generators or by point or line sources such as munitions, covert deposits and moving sprayers.	Biological agents may pose viable threats for hours to years depending on the agent and the conditions in which it exists.	Depending on the agent used and the effectiveness with which it is deployed, contamination can be spread via wind and water. Infection can be spread via human or animal vectors.	Altitude of release above ground can affect dispersion; sunlight is destructive to many bacteria and viruses; light to moderate wind will disperse agents but higher winds can break up aerosol clouds; the micro-meteorological effects of buildings and terrain can influence aerosolization and travel of agents.

Source: FEMA   (*Continued*)

Bioterrorism: A Guide for Facility Managers

Table 9-2. Event Profiles for Terrorism (*Cont'd*)

Hazard	Application Mode	Hazard Duration	Extent of Effects; Static/Dynamic	Mitigating and Exacerbating Conditions
Cyber-terrorism	Electronic attack using one computer system against another.	Minutes to days.	Generally no direct effects on built environment.	Inadequate security can facilitate access to critical computer systems, allowing them to be used to conduct attacks.
Agriterrorism	Direct, generally covert contamination of food supplies or introduction of pests and/or disease agents to crops and livestock.	Days to months.	Varies by type of incident. Food contamination events may be limited to discrete distribution sites, whereas pests and diseases may spread widely. Generally no effects on built environment.	Inadequate security can facilitate adulteration of food and introduction of pests and disease agents to crops and livestock.
Radiological Agent	Radioactive contaminants can be dispersed using sprayers/aerosol generators, or by point or line sources such as munitions, covert deposits and moving sprayers.	Contaminants may remain hazardous for seconds to years depending on material used.	Initial effects will be localized to site of attack; depending on meteorological conditions, subsequent behavior of radioactive contaminants may be dynamic.	Duration of exposure, distance from source of radiation, and the amount of shielding between source and target determine exposure to radiation.
Nuclear Bomb	Detonation of nuclear device underground, at the surface, in the air or at high altitude.	Light/heat flash and blast/shock wave last for seconds; nuclear radiation and fallout hazards can persist for years. Electromagnetic pulse from a high-altitude detonation lasts for seconds and affects only unprotected electronic systems.	Initial light, heat and blast effects of a subsurface, ground or air burst are static and are determined by the device's characteristics and employment; fallout of radioactive contaminants may be dynamic, depending on meteorological conditions.	Harmful effects of radiation can be reduced by minimizing the time of exposure. Light, heat and blast energy decrease logarithmically as a function of distance from seat of blast. Terrain, forestation, structures, etc. can provide shielding by absorbing and/or deflecting radiation and radioactive contaminants.
Hazardous Material Release (fixed facility or trans-portation)	Solid, liquid and/or gaseous contaminants may be released from fixed or mobile containers.	Hours to days.	Chemicals may be corrosive or otherwise damaging over time. Explosion and/or fire may be subsequent. Contamination may be carried out of the incident area by persons, vehicles, water and wind.	As with chemical weapons, weather conditions will directly affect how the hazard develops. The micro-meteorological effects of buildings and terrain can alter travel and duration of agents. Shielding in the form of sheltering in place can protect people and property from harmful effects. Non-compliance with fire and building codes as well as failure to maintain existing fire protection and containment features can substantially increase the damage from a hazardous materials release.

Source: FEMA

Expanding the Asset Inventory

Once the initial asset inventory has been completed, the list should be expanded to include other critical facilities, sites, systems or locations within the area. These other facilities within the area could potentially be targeted for attack. And, depending upon the method and nature of the attack, the impact may spread well beyond the location of origin. Table 9-3, designed as an aid to identifying other critical facilities, sites, systems and area assets, follows.

ASSESSING VULNERABILITIES

The vulnerabilities of a given facility, site, system or other asset can be identified as either an *inherent vulnerability* or a *tactical vulnerability*.

Inherent vulnerabilities exist in any given place in the environment that is independent of any applied protective or mitigating actions. Stadiums, convention centers, etc., are places where thousands of people gather and may be attractive targets to a terrorist. An assessment of such inherent vulnerabilities must be conducted for each asset to determine its weaknesses. Using the asset inventory described above, an assessment of the inherent vulnerability of each asset can be based on the following factors:

- *Visibility.* What is the awareness level of the existence of the facility, site, system or location?

- *Utility.* What value would potential terrorists place upon the facility, site, system or location, in terms of meeting their objectives?

- *Accessibility.* How accessible is the facility, site, system or location?

- *Asset mobility.* Is the asset's location fixed or mobile? If mobile, how often is it moved, relocated or repositioned?

- *Presence of hazardous materials.* Are flammable, explosive, biological, chemical and/or radiological materials stored on site?

- *Potential for collateral damage.* What are the potential consequences for the surrounding area if the facility, site, asset or location is attacked or damaged?

Table 9-3. Asset Identification Checklist

Local, state and federal government offices
(list all in your jurisdiction)

☐ _____

☐ _____

☐ _____

☐ _____

Military installations, including Reserve and National Guard component facilities (list all in your jurisdiction)

☐ _____

☐ _____

☐ _____

☐ _____

Emergency services

☐ Backup facilities

☐ Communication centers

☐ Emergency operations centers

☐ Fire/Emergency Medical Service (EMS) facilities

☐ Law enforcement facilities

Politically or symbolically significant sites

☐ Embassies, consulates

☐ Landmarks, monuments

☐ Political party and special interest group offices

☐ Religious sites

Transportation infrastructure components

☐ Airports

☐ Bus stations

☐ Ferry terminals

☐ Interstate highways

☐ Oil/gas pipelines

☐ Railheads/rail yards

☐ Seaports/river ports

☐ Subways

☐ Truck terminals

☐ Tunnels/bridges

Energy, water, and related utility systems

☐ Electricity production, transmission, and distribution system components

☐ Oil and gas storage/shipment facilities

☐ Power plant fuel distribution, delivery, and storage

☐ Telecommunications facilities

☐ Wastewater treatment plants

☐ Water supply/purification/distribution systems

Telecommunications and information systems

☐ Cable TV facilities

☐ Cellular network facilities

☐ Critical cable routes

☐ Major rights of way

☐ Newspaper offices and production/distribution facilities

☐ Radio stations

☐ Satellite base stations

☐ Telephone trunking and switching stations

☐ Television broadcast stations

Health care system components

☐ Emergency medical centers

☐ Family planning clinics

☐ Health department offices

☐ Hospitals

☐ Radiological material and medical waste transportation, storage, and disposal

☐ Research facilities, laboratories

☐ Walk-in clinics

Source: FEMA

Table 9-3. Asset Identification Checklist (*Cont'd*)

Financial services infrastructures and institutions

☐ Armored car services

☐ Banks and credit unions

Agricultural facilities

☐ Chemical distribution, storage, and application sites

☐ Crop spraying services

☐ Farms and ranches

☐ Food processing, storage, and distribution facilities

Commercial/manufacturing/industrial facilities

☐ Apartment buildings

☐ Business/corporate centers

☐ Chemical plants (include facilities having Section 302 Extremely Hazardous Substances on-site)

☐ Factories

☐ Fuel production, distribution, and storage facilities

☐ Hotels and convention centers

☐ Industrial plants

☐ Malls and shopping centers

☐ Raw material production, distribution, and storage facilities

☐ Research facilities, laboratories

☐ Shipping, warehousing, transfer, and logistical centers

Mobile assets

☐ Aviation and marine units

☐ Mobile emergency operations centers/command centers

☐ Portable telecommunications equipment

☐ Red Cross Emergency Response Vehicles, Salvation Army mobile canteens, etc.

☐ Other (Bloodmobiles, mobile health clinics, etc.)

Recreational facilities

☐ Auditoriums

☐ Casinos

☐ Concert halls and pavilions

☐ Parks

☐ Restaurants and clubs frequented by potential target populations

☐ Sports arenas and stadiums

☐ Theaters

Public/private institutions

☐ Academic institutions

☐ Cultural centers

☐ Libraries

☐ Museums

☐ Research facilities, laboratories

Events and attractions

☐ Festivals and celebrations

☐ Open-air markets

☐ Parades

☐ Rallies, demonstrations, and marches

☐ Religious services

☐ Scenic tours

☐ Theme parks

Source: FEMA

- *Occupancy*. What is the potential for mass casualties based on the maximum number of individuals on site at a given time?

The following table (Table 9-4) is designed to serve as an aid in determining how vulnerable each identified asset is, and how vulnerable the assets are—relative to each other.

Tactical Vulnerabilities of assets are based on several factors. These factors are:

- Site perimeter
  Site planning/design: Is the facility designed with security in mind— both site-specific and with regard to adjacent land uses? Parking security: Are vehicle access and parking managed in a way that separates vehicles and structures?

- Building envelope
  Structural engineering: Is the building's envelope designed to be blast-resistant? Does it provide protection against biological chemical and radiological contaminants?

- Facility interior
  Architectural and interior space planning: Does security screening cover all public and private areas? Are public and private activities separated? Are critical building systems and activities separated?
  Mechanical engineering: Are the utilities and HVAC systems protected and/or backed-up with redundant systems?
  Electrical engineering: Are emergency power and telecommunications available? Are all alarm systems operational? Is lighting sufficient?
  Fire protection: Are the building's water supply and fire suppression systems adequate? Are they compliant with all applicable codes? Are on-site personnel appropriately trained in basic fire prevention techniques? Are local first responders aware of the facility floor plan and business operations?
  Electronic and organized security: Are systems (and personnel) in place to monitor and protect the facility?

MITIGATION PRIORITIES

There are several elements that must be addressed in establishing mitigation priorities. They are: (1) critical assets; (2) vulnerability of the critical assets; and (3) threat.

Table 9-4.

Facility Inherent Vulnerability Assessment Matrix

The Facility Inherent Vulnerability Assessment Matrix *provides a way to record how vulnerable each asset is and enables the planning team to compare how vulnerable the assets are relative to each other. Make a copy for each asset and fill in the facility name or other identifier in the space provided. Select the appropriate point value for each criterion based on the description in each row. Then add the point values to get the total for each asset. When you have done this for each asset you identified, compare the total scores to see how the assets rank in relation to one another.*

Facility _____

Vulnerability Point Values

Criteria	0	1	2	3	4	5	Score
Asset Visibility	–	Existence not well known	–	Existence locally known	–	Existence widely known	
Target Utility	None	Very Low	Low	Medium	High	Very High	
Asset Accessibility	Remote location, secure perimeter, armed guards, tightly controlled access	Fenced, guarded, controlled access	Controlled access, protected entry	Controlled access, unprotected entry	Open access, restricted parking	Open access, unrestricted parking	
Asset Mobility	–	Moves or is relocated frequently	–	Moves or is relocated occasionally	–	Permanent / fixed in place	
Presence of Hazardous Materials	No hazardous materials present	Limited quantities, materials in secure location	Moderate quantities, strict control features	Large quantities, some control features	Large quantities, minimal control features	Large quantities, accessible to non-staff persons	
Collateral Damage Potential	No risk	Low risk / limited to immediate area	Moderate risk / limited to immediate area	Moderate risk within 1-mile radius	High risk within 1-mile radius	High risk beyond 1-mile radius	
Site Population/ Capacity	0	1-250	251-500	501-1000	1001-5000	> 5000	
						TOTAL	

Source: FEMA

Critical Assets

A critical asset is a measure of the importance of the facility or system to the community. To determine if the asset is critical, consider the following:

•   Is it an element of one of the community's critical infrastructures?

•   Does it play a key role in the community, economy or culture?

•   What are the consequences of destruction, failure, or loss of function of the asset in terms of fatalities and/or injuries, property losses, and economic impact?

•   What is the likelihood of subsequent consequences should the asset be destroyed or its function lost?

Vulnerability

By identifying the most exploitable weaknesses of each critical asset, the planning team can identify vulnerabilities in greatest need of attention. This, in effect, gives the planning team a criterion to use in establishing mitigation priorities so that the community can focus its efforts on addressing the most critical issues.

Threat

While difficult to quantify, threat is fundamental to the mitigation planning process. The types of unwanted events, i.e., attacks, etc., that could occur depend upon the nature of the business operation, facility population, or proximity to other target areas. For example, a company located near an airport, nuclear energy facility, or a major highway, etc., could experience the effects of an attack upon its neighboring facility.

For mitigation planning purposes, then, planning should focus on the types of incidents most likely to occur.

Estimating Losses

When assessing risk the potential losses from both natural and man-made hazards are generally grouped into the following three categories:

•   People—death and injury;
•   Assets—structures and their contents; and
•   Functions—provision of services and generation of revenue.

Terrorism and technological occurrences, however, present different implications for loss estimation. Because of the unpredictability of human nature, it is difficult to determine when and if an attack would occur.

In some cases, worse case scenarios could be developed and losses estimated if the hazard can be characterized with some accuracy. For example, models can be used to estimate the consequences of various chemical release scenarios. This can be done by using the location of the rail transport lines and the kinds and quantities of the hazardous chemicals transported over them. Particular attention can be paid to evacuating occupants. When warranted, attention should also be paid to wind direction and stream flow. Both are geologic elements that can disperse contaminants beyond the perimeter of the incident.

When addressing human-engineered events such as bombs, incendiary devices, etc., it is important to remember that these are difficult to quantify. As such, the planning team might consider worst-case scenarios.

DEVELOPING THE PLAN

There are a number of steps involved in developing the mitigation plan. These steps are:
Step 1: Developing mitigation goals and objectives
Step 2: Identifying, prioritizing and implementing mitigation actions
Step 3: Documenting the mitigation planning process

Step 1: Developing mitigation goals and objectives.
The process for developing goals and objectives for the terrorism mitigation plan is the same as developing mitigation goals and objectives for natural disaster occurrences that may impact a company and its operations. In either case, terrorism mitigation and natural disaster mitigation have the same goals—to ensure occupant safety and minimize injury; to protect property; to reduce disaster response costs; and to ensure business continuity after an occurrence.

After completing the risk assessment and estimating losses, those critical assets with the greatest risk can be identified. This information should be incorporated into the company's overall disaster and recovery plan. By combining this information with the applicable data for natural disasters, facility managers and building owners will have a comprehensive understanding of critical asset vulnerabilities.

Step 2: Identifying, prioritizing and implementing mitigation actions
Once mitigation goals and objectives have been developed, the specific actions that will be taken to achieve them should be identified. It is important to note that some of the actions taken to mitigate natural hazards can also provide protection against human-engineered occurrences. For example, fire mitigation techniques may help protect facilities against the effects of bombs and incendiary attacks. Examples of such techniques include improved sprinkler systems, increased use of fireproofing and/or fire-resistant materials, redundant water supplies for fire protection and site setbacks.

High wind mitigation techniques that provide building envelope protection and structural strengthening may also help mitigate against impact or explosion effects of bombs. Examples of such techniques include openings that use windows with impact-resistant laminated glazing, improving connections and the local path of the building, and adding and/or reinforcing shear walls.

Earthquake mitigation techniques that provide for structural strengthening of buildings may help resist the impact and/or explosion effects of bombs. Examples of such techniques include adding steel moment frames, shear walls, cross bracing, stronger floor systems, walls reinforced with shotcrete/fiber materials, columns reinforced with fiber wraps or steel jackets, tension/shear anchors, vibration dampers, and strengthening or providing additional detailing of the building's connections.

While the mitigation goals for both natural and human-engineered occurrences are the same, there is a difference in the mitigation strategies to be employed. Mitigation strategies for human-engineered occurrences focus primarily on three elements. These elements are (1) to create environments that are difficult to attack; (2) that are resilient to the consequences of the attack; and (3) most importantly, mitigation strategies are developed to protect a facility's occupants, should an attack occur. For example, mitigation strategies could involve "target hardening" actions. These target hardening actions range from relatively small-scale projects such as installation of security fencing around the building's HVAC system's air intake, to larger-scale projects such as landscape redesign. Table 9-5 lists potential human-engineered mitigation actions.

Table 9-5. Terrorism and Technological Hazard Mitigation Actions

Terrorism and Technological Hazard Mitigation Actions

The list of actions below is by no means exhaustive or definitive; rather, it is intended as a point of departure for identifying potential mitigation techniques and strategies in your community or state.

Site Planning and Landscape Design

- Implement Crime Prevention Through Environmental Design (CPTED)
- Minimize concealment opportunities in landscaping and street furniture, such as hedges, bus shelters, benches, and trash receptacles
- Design grounds and parking facilities for natural surveillance by concentrating pedestrian activity, limiting entrances/exits, and eliminating concealment opportunities
- Separate vehicle and pedestrian traffic
- Implement vehicle and pedestrian access control and inspection at perimeter (ensure ability to regulate flow of people and vehicles one at a time)
- Design site circulation to minimize vehicle speeds and eliminate direct approaches to structures
- Incorporate vehicle barriers such as walls, fences, trenches, ponds/basins, plantings, trees, sculptures, and fountains into site planning and design
- Ensure adequate site lighting
- Design signage for simplicity and clarity
- Locate critical offices away from uncontrolled public areas
- Separate delivery processing facilities from remaining buildings
- Maintain access for emergency responders, including large fire apparatus
- Identify and provide alternate water supplies for fire suppression
- Eliminate potential site access through utility tunnels, corridors, manholes, etc.

Architectural and Interior Space Planning

- Collocate/combine staff and visitor entrances; minimize queuing in unprotected areas
- Incorporate employee and visitor screening areas into planning and design
- Minimize device concealment opportunities such as mailboxes and trash receptacles outside screened areas
- Prohibit retail activities in non-secured areas

- Do not locate toilets and service spaces in non-secured areas
- Locate critical assets (people, activities, systems) away from entrances, vehicle circulation and parking, and loading and maintenance areas
- Separate high-risk and low-risk activities
- Separate high-risk activities from areas accessible to the public
- Separate visitor activities from daily activities
- Separate building utilities from service docks, and harden utilities
- Locate delivery and mail processing facilities remotely or at exterior of building; prevent vehicles from driving into or under building
- Establish areas of refuge; ensure that egress pathways are hardened and discharge into safe areas
- Locate emergency stairwells and systems away from high-risk areas
- Restrict roof access
- Ensure that walls, doors, windows, ceilings, and floors can resist forced entry
- Provide fire- and blast-resistant separation for sprinkler/standpipe interior controls (risers) and key fire alarm system components
- Use visually open (impact-resistant, laminated glass) stair towers and elevators in parking facilities
- Design finishes and signage for visual simplicity

Structural Engineering

- Create blast-resistant exterior envelope
- Ensure that structural elements can resist blast loads and progressive collapse
- Install blast-resistant exterior window systems (frames, security films, and blast curtains)
- Ensure that other openings (vents, etc.) are secure and blast-resistant
- Ensure that mailrooms are secure and blast-resistant
- Enclose critical building components within hardened walls, floors, and ceilings

(continued)

Source: FEMA

Table 9-5. Terrorism and Technological Hazard Mitigation Actions (*Cont'd.*)

Mechanical Engineering

- Locate utility and ventilation systems away from entrances, vehicle circulation and parking, and loading and maintenance areas
- Protect utility lifelines (water, power, communications, etc.) by concealing, burying, or encasing
- Locate air intakes on roof or as high as possible; if not elevated, secure within CPTED-compliant fencing or enclosure
- Use motorized dampers to close air intakes when not operational
- Locate roof-mounted equipment away from building perimeter
- Ensure that stairways maintain positive pressure
- Provide redundant utility and ventilation systems
- Provide filtration of intake air
- Provide secure alternate drinking water supply

Electrical Engineering

- Locate utility systems and lifelines away from entrances, vehicle circulation and parking, and loading and maintenance areas
- Implement separate emergency and normal power systems; ensure that backup power systems are periodically tested under load
- Locate primary and backup fuel supplies away from entrances, vehicle circulation and parking, and loading and maintenance areas
- Secure primary and backup fuel supply areas
- Install exterior connection for emergency power
- Install adequate site lighting
- Maintain stairway and exit sign lighting
- Provide redundant telephone service
- Ensure that critical systems are not collocated in conduits, panels, or risers
- Use closed-circuit television (CCTV) security system

Fire Protection Engineering

- Ensure compliance with codes and standards, including installation of up-to-date fire alarm and suppression systems
- Locate fire protection water supply system critical components away from entrances, vehicle circulation and parking, and loading and maintenance areas
- Identify/establish secondary fire protection water supply
- Install redundant fire water pumps (e.g., one electric, one diesel); locate apart from each other
- Ensure adequate, redundant sprinkler and standpipe connections
- Install fire hydrant and water supply connections near sprinkler/standpipe connections
- Supervise or secure standpipes, water supply control valves, and other system components

- Implement fire detection and communication systems
- Implement redundant off-premises fire alarm reporting
- Locate critical documents and control systems in a secure yet accessible place
- Provide keybox near critical entrances for secure fire access
- Provide fire- and blast-resistant fire command center
- Locate hazardous materials storage, use, and handling away from other activities
- Implement smoke control systems
- Install fire dampers at fire barriers
- Maintain access to fire hydrants
- Maintain fire wall and fire door integrity
- Develop and maintain comprehensive pre-incident and recovery plans
- Implement guard and employee training
- Conduct regular evacuation and security drills
- Regularly evaluate fire protection equipment readiness/adequacy

Security

- Develop backup control center capabilities
- Secure electrical utility closets, mechanical rooms, and telephone closets
- Do not collocate security system wiring with electrical and other service systems
- Implement elevator recall capability and elevator emergency message capability
- Implement intrusion detection systems; provide 24-hour off-site monitoring
- Implement and monitor interior boundary penetration sensors
- Implement color closed-circuit television (CCTV) security system with recording capability
- Install call boxes and duress alarms
- Install public and employee screening systems (metal detectors, x-ray machines, or search stations)

Parking

- Minimize off-site parking on adjacent streets/lots and along perimeter
- Control all on-site parking with ID checks, security personnel, and access systems
- Separate employee and visitor parking
- Eliminate internal building parking
- Ensure natural surveillance by concentrating pedestrian activity, limiting entrances/exits, and eliminating concealment opportunities
- Use transparent/non-opaque walls whenever possible
- Prevent pedestrian access to parking areas other than via established entrances

Source: FEMA

Prioritizing and Implementing Actions

A cost-benefit analysis should be conducted for any mitigation action that is being considered. Factors to consider when prioritizing hazard mitigation actions include:

• Cost(s);

• Dollar value of risk reduction for each occurrence;

• Frequency with which the benefits of the action will be realized; and

• Time value of money—i.e., present v. future worth.

There are several challenges that arise:

1. *The probability of an attack or frequency of the hazard occurrence is not known.* While it is possible to estimate when or how often a natural hazard would occur—flooding from spring thaws, etc., it is difficult to predict the likelihood of a terrorist attack or a technological occurrence. Therefore, a qualitative approach should be considered. This approach is based upon critical asset threat, and vulnerability of the critical asset itself. In other words, it is important to estimate the likelihood of an attack rather than project the timing of the occurrence.

2. *The deterrence rate may be unknown.* Once a facility is hardened, the deterrence or prevention actions taken may force a terrorist to turn to a less protected facility.

3. *The lifespan of the action may be difficult to quantify.* When planning for future benefits of a hazard mitigation action, the successful performance of the action over the course of its useful life must be taken into account. That protective action may be damaged or destroyed in a single manmade attack or accident. For example, blast-resistant window film may be effective in protecting lives from flying glass, but may need to be replaced after one attack.

Incorporating the threat of terrorism and human-engineered hazards into the planning process can reduce building/facility vulnerability to these types of disasters.

DOCUMENT THE MITIGATION PLANNING PROCESS

The mitigation plan for terrorist attacks and human-engineered occurrences is based on the risk assessment. A comprehensive strategy to address mitigation priorities should be integrated into the natural hazard mitigation plan and should include:

- A summary of the planning process
- An outline the sequence of actions to be taken
- A list the planning team members
- Risk assessment results and loss estimation
- Mitigation goals and objectives
- Actions that will meet established mitigation goals and objectives
- Implementation strategies that detail how the mitigation actions will be implemented

The hazard mitigation plan should serve as the focal point and basis for mitigation decisions for human-engineered and natural disasters. (Since a single comprehensive plan is easier to manage and implement than a series of stand-alone documents, all hazards—natural and human-engineered—should be incorporated into one mitigation plan.) As such, the plan should be written so that anyone who reads it can gain an understanding of hazards and risks as well as the intended solutions.

SUMMARY

The basics of mitigating hazards before they become disasters are similar for both natural and human-engineered occurrences. An accident, deliberate attack or natural disaster such as a crippling snow storm, may not always be able to be prevented, but a well planned and effectively implemented mitigation plan will help to reduce the consequences of such incidents. By using a well-developed mitigation plan, facility managers and building owners will be able to identify, prioritize and implement those mitigation actions necessary to ensure the safety of building occupants.

Sources

Federal Emergency Management Agency (FEMA), *Integrating Manmade Hazards Into Mitigation Planning*, FEMA 386-7. September, 2003.
Gustin, Joseph F. *Disaster and Recovery Planning: A Guide for Facility Managers, 3rd ed.*, Lilburn, GA: The Fairmont Press, Inc., 2004.

Chapter 10

Evacuation and Shelter-In-Place

Generally speaking, there are two primary considerations that determine the order to evacuate. These considerations are:

1. The specific emergency occurrence, and
2. The potential threat to occupant safety.

As such, a facility's evacuation plan must then identify when and how employees, occupants and visitors are to respond to the different types of emergencies. Therefore, occupants need to be trained to respond differently to different threats.

For example, if an airborne hazard is detected inside a building, there are several possible actions that can be implemented—singly or in combination—for relatively short periods when a hazard becomes apparent. These measures, which include sheltering-in-place, using protective masks, evacuating the building, and purging with smoke fans, apply only to chemicals that have:

- A warning property (e.g., odor);
- Agents detectable by automatic detectors; or
- In response to the release of chemicals (e.g., an accident at a chemical storage facility or a chemical spill on a highway adjacent to the facility).

To ensure that these actions can be carried out effectively, an action plan that includes occupant training must be specific to the building or facility.

EVACUATION AND AIRBORNE HAZARDS

Evacuation is the most common protective action taken when an airborne hazard, such as smoke or a noxious odor, is detected in the building. In most cases, existing plans for fire evacuation are applicable for evacuation in response to a chemical hazard. Evacuation is the simplest protective action, but it may not be the best course of action for an external release. This is particularly true if the release is widespread. If the area enveloped by the plume of hazardous material is too large to exit rapidly, the use of sheltering-in-place or protective masks should be considered. Two considerations that must be addressed in a non-fire evacuation are:

- Whether the source of the hazard is internal or external; and
- Whether or not the evacuation might lead to other risks.

SHELTERING-IN-PLACE

Sheltering-in-place and evacuation are the protective actions planned for in the event of an accidental release of toxic chemicals. The advantage of sheltering-in-place is that it can be implemented more rapidly than evacuation. The disadvantage is that the protection it provides is variable and diminishes with the duration of the exposure. Sheltering-in-place requires two distinct actions to be taken without delay to maximize the passive protection a building can provide. These actions are:

1. To reduce the air exchange rate of the building before the hazardous plume arrives, by closing all windows and doors and turning off all fans, air conditioners and combustion heaters.

2. To increase the air exchange rate of the building as soon as the hazardous plume has passed, by opening all windows and doors and turning on all fans to aerate the building.

Though tightly sealed, a building does not prevent contaminated air from entering. Rather, it minimizes the rate of infiltration. Outside air enters slowly and once the external hazard has passed, the building releases the contaminated air slowly as long as it remains closed.

The level of protection that can be attained by sheltering-in-place is substantial, but is much less than that of high-efficiency filtering of the fresh air introduced into the building. The amount of protection varies with:

- the building's air exchange rate;
- duration of exposure;
- purging or period of occupancy; and
- natural filtering.

The Building's Air Exchange Rate—the tighter the building and the lower the air exchange rate, the greater the protection it provides. In most cases, air conditioners and combustion heaters cannot be operated while sheltering-in-place. Doing so increases the air exchange rate.

Duration of Exposure—protection varies with time, diminishing as the time of exposure increases. Therefore, sheltering-in-place is suitable only for exposures of short duration.

Purging or Period of Occupancy—the length of time occupants remain in the building after the hazardous cloud has passed also affects the level of protection. Because the building slowly releases contaminants that have entered, at some point during cloud passage the concentration inside exceeds the concentration outside. Increasing the air exchange rate after cloud passage, or exiting the building into clean air, offers maximum protection.

Natural Filtering—natural filtering occurs to a small degree when the hazardous material is deposited in the building shell or on interior surfaces as it passes into and out of the building. The tighter the building, the greater the natural filtering contributes to the protection factor.

Unlike the actions required for sheltering in a home which are relatively simple (i.e., closing windows and doors and turning off all air conditioners, fans and combustion heaters), doing so in a building requires more time and planning. All air handling units must be turned off and dampers, if controllable, must be closed.

Although sheltering is generally for protection against an outdoor release, it is possible, though more complex, to shelter-in-place against an internal release in a large building, i.e., to shelter on one floor when a hazard exists on another floor.

SMOKE PURGE FANS

Activating the facility's smoke purge fans is a means of reducing the hazards to which building occupants are exposed. However, it is mainly useful when the source of the hazard is inside the building. Smoke purge fans can be used to purge the building after an external release, once the outdoor hazard has cleared. Again, the use of smoke purge fans is beneficial for the purging phase of sheltering-in-place, once it has been confirmed that the agent is no longer present outside the building.

PROTECTIVE MASKS

New models of universal-fit escape masks have been developed for protection against chemical and biological agents. Such masks form a seal at the wearer's neck, and therefore, fit a wider range of sizes than the traditional masks that seal around the face. These universal-fit masks do not require special fitting techniques or multiple sizes. They also are designed to store compactly and can be carried on the belt. Their protective capabilities are equivalent to military masks and have a five year shelf life. Some considerations for their use should be noted. Although the filters of these masks are designed for a broad range of chemicals, they do not filter all chemicals. For example, they are not effective against certain toxic chemicals of high vapor pressure, such as carbon monoxide. These masks can be kept at an employee's desk for accessibility in the event of an emergency occurrence or they can be worn on the belt by security personnel.

However, a note of caution is warranted. All protective masks, including respirators, escape hoods, gas masks, etc., require training to be properly used. As such, employees and building occupants must be trained in the proper usage of this equipment. Areas that must be included in the training are proper storage, maintenance, use and discarding. While this information is provided by the supplier of the equipment (i.e., seller, distributor or manufacturer), employees and building occupants must receive actual hands-on training in proper procedures and techniques.

EVACUATION

As previously noted, there are two primary considerations that will impact the decision to evacuate—the specific emergency occurrence, itself, and the potential threat that the occurrence presents to occupant safety.

A disorganized evacuation can result in confusion, injury and/or property damage. When developing evacuation plans it is important to determine the following:

- Conditions under which an evacuation would be necessary
- Conditions under which it may be better to shelter-in-place
- A clear chain of command and designation of the person authorized to order the evacuation (or building shutdown)
- Specific evacuation procedures, including routes and exits
- Specific evacuation procedures for high-rise buildings
  — For employers
  — For employees
- Procedures for assisting employees, occupants and visitors to evacuate, particularly those with disabilities or who do not speak English
- Designation of what, if any, employees will remain after the evacuation order to shut down critical operations or perform other duties before evacuating
- A means of accounting for employees and occupants after an evacuation
- Special equipment for employees
- Appropriate respirators

Conditions for an Evacuation

There are a number of emergencies, both human engineered and natural, that may require a building or facility to be evacuated. These emergency occurrences include:

- Chemical, biological and radiological releases
- Fires
- Explosions
- Civil disturbances
- Workplace violence
- Floods
- Earthquakes
- Hurricanes

Routes and Exits

An orderly and complete evacuation of all occupants and visitors requires a careful provision for egress routes and accounting for all individuals after the evacuation. Most facilities will have maps or floor diagrams with arrows that designate the exit route assignments. These maps should include locations of exits, assembly points, and equipment (such as fire extinguishers, first aid kits, spill kits, etc.) that may be needed in an emergency. Exit routes should be:

• Clearly marked and well-lit;
• Wide enough to accommodate the number of evacuating personnel;
• Unobstructed and clear of debris at all times; and
• Unlikely to expose evacuating personnel to additional hazards.

These maps and / or floor diagrams that show evacuation routes and exits must be posted prominently for all occupants to see.

Means of Egress
As defined in the Life Safety Code Handbook published by the National Fire Protection Association (NFPA), means of egress is the continuous and unobstructed way of travel from any point in a building or structure to a public way consisting of three separate and distinct components. These components are:

• Exit access
• Exit
• Exit discharge

Exit Access—The exit access is that component of a means of egress that leads to an exit. For example, an exit access includes rooms and the building spaces that people occupy, as well as doors, corridors, aisles, unenclosed stairs and enclosed ramps that must be traveled to reach an exit.
Exit—An exit is that component of the means of egress that is separated from all other spaces of a building by either construction—which must have the minimum degree of fire resistance—or equipment so that a protected way of travel to the point of exit discharge is provided. Exits include exterior exit doors, exit passageways, separated exit stairs and separated exit ramps.
Exit Discharge—The third component that comprises a means of egress is the exit discharge, or the path of travel from the termination of an exit to a public way. Since some building exits do not discharge directly into a public way, the path of travel may be within the building itself or outside the building. In either case, the purpose is to provide

building occupants with the means for reaching safety.

Each of these components comprises the means of egress which must be provided from every location within the facility. Accordingly, all means of egress within the facility must be routinely inspected and regularly maintained to ensure maximum operability and utility by occupants seeking safety.

Types of Evacuation

There are two types of evacuation—partial evacuation and total, or complete evacuation. While the nature of the disaster/emergency, as well as the potential threat to the safety of building occupants determines the type of evacuation undertaken, the *immediacy* of the threat must also be considered.

Partial Evacuation—In a partial evacuation, occupants who are immediately affected, or whose safety may be jeopardized by the occurrence, are relocated from the endangered area to a safe or secured area. This secured area may be either located within the facility itself or away from the building. For example, when fire is detected in a multistory facility or high-rise building, occupants on the floor immediately above and below the origin of the fire are evacuated.

Total Evacuation—In a total or complete evacuation, all occupants are required to vacate the site premises with the possible exception of the disaster/emergency team members. In some situations, again depending upon the nature and immediacy of the disaster or emergency, team members may remain behind to ensure that all other occupants have vacated the site and/or to secure the building proper and critical building contents.

Evacuation Factors

There are several factors that must be considered in determining which type of evacuation must be undertaken. These factors are:

- The nature of the disaster/emergency occurrence; and

- The potential threat to occupant safety, as determined by the severity of the occurrence.

In turn, each of these factors determines the immediacy of the evacuation need.

Other Evacuation Factors
There are a number of other critical factors related to evacuation that must be addressed in developing evacuation strategies. These factors include determining:

1. Internal authority/responsibility levels for offering and supervising the evacuation; and
2. Training.

Authority Levels—when an evacuation is necessary, responsible and trained individuals are needed. These trained individuals should coordinate activities to ensure a safe and successful evacuation

It is a common practice to select a responsible individual to lead and coordinate evacuation. It is critical that employees and building occupants know who the coordinator is and understand that this person has the authority to make decisions during an emergency occurrence.

The designated coordinator assumes responsibility for assessing a situation to determine whether an occurrence requires activation of emergency procedures, notifying and coordinating with outside emergency services, and directing shut down of utilities or building operations if necessary.

In other instances local emergency officials, such as the local fire department, may order evacuation of the building or facility. If access to radio or television is available, news broadcasts will provide up to date information.

When emergency officials, such as the local fire department, respond to an emergency at the facility, they will assume responsibility for the safety of building occupants. They also have the authority to make decisions regarding evacuation and any and all other actions that they deem necessary to protect life and property. Depending upon the potential threat to occupant safety, any decision to evacuate may need to be made immediately.

Training—enough people must be designated and trained in order to assist in the safe and orderly emergency evacuation of employees and building occupants. In fact, OSHA stipulates such in their outline for developing an emergency action plan. [OSHA standard 1910.38(a)(5)(i)]. Training should be offered to employees when the initial emergency action plan is developed and to all newly hired employees [OSHA standard 1910.38(5)(ii)(A)]. Employees should be retrained in emergency evacuation procedures when their actions or responsibilities under the plan change [OSHA standard 1910.38(a)(5)(ii)(B)]. These employees should also be retrained when the plan, itself, changes due to a change in the layout or design of the facility, new equipment, hazardous materials or processes are introduced that affect evacuation routes, or new types of hazards are introduced that require special actions [OSHA standard 1910.38(a)(5)(ii)(C)].

Employees and building occupants should be trained and educated in the types of emergencies that may occur. They also need to be trained in the proper course of action to be taken. Facility size, number of occupants, processes used, materials handled, and the availability of onsite or outside resources will determine training requirements. Additionally, designated individuals must be trained in the various types of assistance that will be made available to persons with disabilities. Assistance may come in the form of trained in-house personnel and/or fire and emergency services personnel in assisting in the evacuation of occupants with disabilities.

Employees and building occupants must understand the types of potential emergencies, reporting procedures, alarm systems, evacuation plans and shut down procedures. They should also be made aware of any special hazards that may be stored onsite such as flammable materials, toxic chemicals, radioactive sources or water-reactive substances. The facility manager or building owner must also inform employees of the fire hazards present in the facility [OSHA standard 1910.38(b)(4)(i)]. Employees and building occupants must know who will be in charge during an emergency.

Training for employees should address the following:

- Individual roles and responsibilities
- Threats, hazards and protective actions
- Notification, warning and communication procedures
- Means for locating family members in an emergency
- Emergency response procedures
- Evacuation, shelter and accountability procedures
- Location and use of common emergency equipment
- Emergency shutdown procedures

It is important to keep in mind that if training is not reinforced it will be forgotten. Retraining should be done on an annual basis, minimally, and practice sessions should be held periodically. Practice drills or sessions should include outside resources such as local fire and police departments when possible. After each drill, management and employees should evaluate the effectiveness of the drill. Both strengths and weaknesses can then be assessed.

As part of the emergency action planning process, the plan should be posted in an area where employees and occupants will have access to it. Each employee/occupant should be familiar with the elements of the plan that dictates personal safety procedures in the event of an emergency. The written emergency action plan detailing evacuation procedures must be available to employees and building occupants and be kept at the facility. If companies have 10 or fewer employees, the plan may be communicated orally, i.e., written plans do not need to be maintained [OSHA standards 1910.38(b)(4)(ii) and 1910.38(a)(5)(iii)].

The emergency action plans, along with evacuation procedures, should be reviewed with other companies or employee groups in multi-tenanted facilities to ensure coordination of effort. In addition, emergency plans should be reviewed with local emergency responders such as the fire department, local HAZMAT teams, or other outside responders. This coordination ensures that facility managers and building owners are aware of the capabilities of these outside responders and that the outside responders will be aware of facility management expectations.

As noted previously, it is important to hold practice evacuation drills. Evacuation drills permit employees and building occupants to become familiar with emergency procedures, egress routes, and assembly locations. Should an actual emergency occur, all employees and building occupants will know proper response procedures and act accordingly. Drills should be conducted as often as necessary to keep everyone prepared.

Facility operations and personnel can change frequently and often do. As such, emergency plans should be regularly updated whenever there is a change in the layout or design of the facility, new equipment, hazardous materials, or processes that affect evacuation routes. Updating should also occur when new types of hazards are introduced that require special actions. Emergency plans should also be updated when a person's emergency actions or responsibilities change.

The most common outdated item in emergency plans is the facility and agency contact information. This information should be located at the front of the emergency plan so that it can easily be updated.

ADDITIONAL CONSIDERATIONS

There are several other considerations regarding evacuation that must be addressed in the planning process. These include the location to which occupants will report upon completion of the partial or total evacuation; confirmation of successful evacuation; procedures for evacuating occupants with special needs; sheltering-in-place and maintenance of evacuation routes.

Reporting Locations

Regardless of the type of evacuation that may be ordered, a reporting location must be identified. In those situations that warrant a partial evacuation, consideration must be given to which area within the facility itself will be made available for evacuated occupants. Areas within the facility that have controlled or limited access for security reasons are not local choices. Neither are any business operations areas that may be disturbed by a sudden influx of additional personnel. The evacuation strategy must address location issues so that unnecessary problems do not crop up during an actual occurrence.

The evacuation strategy must also address the issue of total building evacuation. An off-site location must be identified. The location itself should provide safety from the disaster site and should be large enough to accommodate the total number of evacuees. Without an identified destination, the safety of evacuees could be jeopardized. Additionally, the effectiveness of fire and emergency rescue services could be compromised by large numbers of people congregating outside the disaster site.

Confirming Evacuation

Essentially, there are two methods by which an evacuation can be confirmed—the roll-call (or head count) method, and the search method. Each has its own distinct advantages depending upon occupancy type and facility size.

The Roll-Call Method—For those facilities that house a relatively small number of regular occupants, the roll-call method is an effective means for verifying evacuation. The evacuation supervisor, or designee, can either count heads against the occupant list, or require a voice count against the occupant list to ensure that all occupants have successfully evacuated.

The Search Method—In larger facilities that house a relatively large number of occupants, the roll-call method is often not practical. In a total evacuation of these facilities, wherein large numbers of regular occupants are either tenanted, or in those occupant types that include transients—i.e., mercantile, assembly occupancies—it becomes almost impossible to account for each person. In these instances, the search method is most effective.

Upon authorization from the appropriate jurisdictional agency to re-enter an evacuated site, the search method can be undertaken to determine successful evacuation. In those instances where authorization to re-enter an evacuated site is not immediately given, the jurisdictional agency may be called upon to conduct the post evacuation search.

Procedures for Assisting Special Needs Occupants

In many cases a person is designated to help move employees from danger to safe areas during an emergency. According to OSHA, one designated person (evacuation warden) for every 20 employees should be adequate. The appropriate number of wardens should be available at all times during a facility's normal hours of operations.

Wardens may be responsible for checking offices, bathrooms and other areas before being the last to exit. They might also be tasked with ensuring that fire doors are closed when exiting. All persons designated to assist in emergency evacuation procedures should be trained in the workplace layout and various alternative escape routes if the primary evacuation route becomes blocked. Persons designated to assist in emergency evacuations should be made aware of:

- occupants with special needs who may require extra assistance
- how to use the "buddy system"
- which areas to avoid during an emergency evacuation.

Visitors/transient occupants should also be accounted for following an evacuation and may need additional assistance when exiting. Some facilities have all visitors, contractors and other transient occupants sign in when entering the building and use the list when accounting for all persons in the assembly area. Persons designated to assist in the evacuation are often tasked with helping occupants with special needs to safely evacuate.

In multi-tenanted facilities, it is beneficial for the facility manager to coordinate evacuation plans with all tenant employers. However, OSHA standards do not specifically require this to be done.

Shut Down Procedures: Employees Who Remain Behind and

Building owners and facility managers must review their own specific operations. Doing so will determine whether total and immediate evacuation is possible, as well as when total evacuation is "required" for various types of emergencies. The preferred approach, and the one most often taken by smaller facilities, is the immediate evacuation of all facility occupants when the evacuation order is given, or when the evacuation alarm is sounded.

If there are any employees who are designated to stay behind, the emergency evacuation plan must describe in detail the procedures to be followed by these employees. All individuals remaining behind must be capable of recognizing when to abandon the operation or task and evacuate themselves before their egress path is blocked. In smaller facilities, it is common for the emergency evacuation plan to include the location where utilities can be shut down for all or part of the facility by either the designated person or by emergency response personnel.

Accounting for Occupants

Facility managers and building owners may want to consider including the following steps in the emergency action plan:

- designate assembly areas where employees/occupants should gather after evacuating
- take a head count after the evacuation
- identify the names and last known locations of anyone not accounted for and pass them to the official in charge
- establish a method for accounting for non-employees such as suppliers and customers
- establish procedures for further evacuation in case the incident expands

Shelter-In-Place

Chemical, biological or radiological contaminants may be released into the environment in such quantity and/or proximity to a place of business, that it is safer to remain indoors rather than to evacuate occupants. Such releases may be either intentional or accidental.

"Shelter-in-place" means selecting an interior room(s) that is windowless and taking refuge there. In many cases, local authorities will issue directives to shelter-in-place via television or radio.

Preparing to Evacuate or Stay

Depending upon the circumstances and the type of emergency occurrence, the first decision that building owners and facility managers must make is whether to evacuate or shelter-in-place. In all cases, building owners and facility managers should understand and plan for both possibilities.

In an emergency, local authorities may or may not immediately be able to provide information on what is happening. All available information should be used to assess the situation. If, for example, large amounts of debris are in the air or if local authorities say the air is badly contaminated, sheltering-in-place would be the best available option. However, television or radio broadcasts as well as the internet, should be checked often for new/additional information or special instructions. If the broadcasts indicate that evacuation is required or medical treatment warranted, it should be done immediately.

As part of the emergency evacuation procedure, building owners and facility managers should include the following actions regarding sheltering-in-place:

•   A separate distinct method of alert should be implemented for the shelter-in-place option. A separate, distinct alert method will ensure that employees/occupants shelter-in-place rather than evacuate.

•   Employee/occupants must be trained in shelter-in-place procedures as well as in their roles in implementing the procedures.

Shelter-in-Place Procedures

Specific procedures for shelter-in-place at a facility may include the following:

•   Close the facility.

•   Customers/clients/visitors should be asked to stay, not leave the facility. This will provide for their safety during or subsequent to the occurrence. When authorities provide directions to shelter-in-place, they want everyone to take those steps immediately. Everyone should be instructed not to leave the building.

•   Unless there is imminent threat or danger, employees and occupants, as well as customers, clients, visitors, etc., should call their emergency contacts to let them know where they are and that they are safe.

- Call-forwarding, alternative answering systems or services should be turned on—voice mail recordings should indicate that the business is closed, and that staff and visitors are remaining in the building until building evacuation is authorized.

- Exterior doors should be locked; windows, air vents and dampers should be closed. Fans, heating and air conditioning systems should be turned off; some systems automatically provide for exchange of inside air with outside air; these systems, in particular, should be turned off, sealed or disabled.

- If there is danger of explosion, shades, blinds or curtains should be closed.

- Interior rooms above the ground floor with the fewest windows or vents should be selected. The room(s) should have adequate space for everyone to be able to sit. Overcrowding can be avoided by selecting several rooms. Large storage closets, utility rooms, pantries, copy and conference rooms without exterior windows should be considered. Rooms with mechanical equipment like ventilation blowers or pipes should be avoided because this equipment may not be able to be sealed from the outdoors.

- Essential disaster supplies such as non-perishable food, bottled water, battery-powered radios, first-aid supplies, flashlights, batteries, duct tape, plastic sheeting and plastic garbage bags should be gathered.

- Provisions should be made for hard-wired telephones in the room(s) selected so that emergency contacts can be called and the phones are available if life-threatening conditions need to be reported. It is important to note that cellular telephone equipment may be overwhelmed or damaged during an emergency occurrence.

- Emergency supplies should be moved in the room(s) or into the designated room(s). All windows, doors and vents should be sealed with plastic sheeting and duct tape.

•   The plastic sheeting to be used for sealing windows, doors and air vents should be precut. Each piece should be several inches larger than the space to be covered so that it lies flat against the wall. This will help to prevent contaminated air from entering the room. Label each piece with the location of where it fits.

•   The names of everyone in the room(s) along with their affiliation with the business, i.e., employee, visitor, client, customer, etc., should be written down and reported to emergency personnel.

•   Television and the internet (both, if available) are sources of further information. Local officials may call for evacuation to other areas or may give the "all clear."

Maintaining the Evacuation Route

A final, but no less important consideration is the issue of evacuation route maintenance. As part of a facility's routine inspection and maintenance program, special attention must be given to designated evacuation routes.

Essential from a regulatory standpoint, all routes designated and used in evacuation must be properly maintained. Additionally, all building components must be operable and ready for use in the event that an order to evacuate becomes necessary. These building components include lighting and emergency lighting systems, elevators, smoke and fire resistant ceiling and walls, and stair pressurization systems. And, finally, as previously discussed, all components that comprise the means of egress—the exit access, exit and exit discharge—must be properly maintained so that their utility can be ensured and that occupants can reach a point of safety.

Sources

Cote, Ron, PE, Editor, *Life Safety Code Handbook*, 6th Edition, National Fire Protection Association, Quincy, MA, 1994.

Gustin, Joseph F. *Disaster and Recovery Planning: A Guide for Facility Managers, 3rd ed.*, Lilburn, GA: The Fairmont Press, Inc., 2004.

The Occupational Safety And Health Act of 1970 (December 29, 1970), PL91-596,29 USC 651.

United States Department of The Treasury, Bureau of Alcohol, Tobacco And Firearms, *Bomb And Physical Security Planning*, ATF P 7550.2 (7087).

Appendix I

Building Vulnerability Assessment Screening

A ppendix I provides a tool for a comprehensive assessment of terrorism vulnerability in buildings. It contains a list of questions that provide the basis for identifying the physical and operational vulnerability of the building.

This assessment includes the questions listed Appendix II. Note that Appendix II also provides extended guidance and observations regarding the use of each question in assessing building vulnerability to a terrorist attack.

Table A: Site

Item	Vulnerability Question	Vulnerability estimate	Detailed assessment	Visual inspection	Document review	Org/Mgmt procedure	Moving vehicle	Stationary vehicle	Covert entry	Mail	Supplies	Blast effects	Airborne (contamination)	Waterborne (contamination)
		Characterization					**Terrorist Tactics**							
		Type		**Collection**			**Delivery Methods**					**Mechanisms**		
Site.1a	What major structures surround the facility?	●		●	●									
Site.1b	What critical infrastructure, government, military, or recreation facilities are in the local area that impact transportation, utilities, and collateral damage (attack at this facility impacting the other major structures or attack on the major structures impacting this facility)?	●		●	●	●								
Site.1c	What are the adjacent land uses immediately outside the perimeter of this facility?	●		●	●									
Site.1d	Do future development plans change these land uses outside the facility perimeter?		●	●	●									
Site.2	Does the terrain place the building in a depression or low area?		●	●										
Site.3	In dense, urban areas, does curb lane parking place uncontrolled parked vehicles unacceptably close to a facility in public rights-of-way?	●	●		●			●			●			
Site.4	Is a perimeter fence or other types of barrier controls in place?	●	●					●		●				
Site.5	What are the site access points to the facility?	●		●				●		●				
Site.6	Is vehicle traffic separated from pedestrian traffic on the site?	●	●		●		●		●					
Site.7	Is there vehicle and pedestrian access control at the perimeter of the site?	●	●		●		●		●					
Site.8a	Is there space for inspection at the curb line or outside the protected perimeter?	●	●						●					
Site.8b	What is the minimum distance from the inspection location to the building?	●		●			●		●			●		
Site.9	Is there any potential access to the site or facility through utility paths or water runoff?	●		●	●				●					
Site.10a	What are the existing types of vehicle anti-ram devices for the facility?	●	●				●							
Site.10b	Are these devices at the property boundary or at the building?		●	●			●							
Site.11	What is the anti-ram buffer zone standoff distance from the building to unscreened vehicles or parking?	●		●			●					●		

Source: Federal Emergency Management Agency (FEMA)

Table A: Site (continued)

Item	Vulnerability Question	Vulnerability estimate	Detailed assessment	Visual inspection	Document review	Org/Mgmt procedure	Moving vehicle	Stationary vehicle	Covert entry	Mail	Supplies	Blast effects	Airborne (contamination)	Waterborne (contamination)
Site.12a	Are perimeter barriers capable of stopping vehicles?	●		●	●		●							
Site.12b	Will the perimeter and facility barriers for protection against vehicles maintain access for emergency responders, including large fire apparatus?	●			●									
Site.13	Does site circulation prevent high-speed approaches by vehicles?	●	●				●							
Site.14	Are there offsetting vehicle entrances from the direction of a vehicle's approach to force a reduction of speed?	●	●				●							
Site.15	Is there a minimum setback distance between the building and parked vehicles?	●	●					●				●		
Site.16	Does adjacent surface parking maintain a minimum standoff distance?	●	●					●				●		
Site.17	Do stand-alone, above ground parking facilities provide adequate visibility across as well as into and out of the parking facility?	●	●					●	●					
Site.18	Are garage or service area entrances for employee-permitted vehicles protected by suitable anti-ram devices?	●	●						●	●				
Site.19	Do site landscaping and street furniture provide hiding places?	●	●						●					
Site.20a	Is the site lighting adequate from a security perspective in roadway access and parking areas?	●	●						●					
Site.20b	Are line-of-sight perspectives from outside the secured boundary to the building and on the property along pedestrian and vehicle routes integrated with landscaping and green space?	●	●		●				●					
Site.21	Do signs provide control of vehicles and people?	●	●											
Site.22	Are all existing fire hydrants on the site accessible?	●	●											

Source: Federal Emergency Management Agency (FEMA)

Table B: Architectural

Item	Vulnerability Question	Vulnerability estimate	Detailed assessment	Visual inspection	Document review	Org/Mgmt procedure	Moving vehicle	Stationary vehicle	Covert entry	Mail	Supplies	Blast effects	Airborne (contamination)	Waterborne (contamination)
Arch.1	Does the site and architectural design incorporate strategies from a Crime Prevention Through Environmental Design (CPTED) perspective?	•	•				•	•						
Arch.2	Is it a mixed-tenant facility?	•		•	•	•								
Arch.3	Are pedestrian paths planned to concentrate activity to aid in detection?	•	•						•					
Arch.4	Are there trash receptacles and mailboxes in close proximity to the facility that can be used to hide explosive devices?	•	•										•	
Arch.5	Do entrances avoid significant queuing?	•		•		•								
Arch.6a	Does security screening cover all public and private areas?	•	•			•		•						
Arch.6b	Are public and private activities separated?	•		•		•								
Arch.6c	Are public toilets, service spaces, or access to stairs or elevators located in any non-secure areas, including the queuing area before screening at the public entrance?	•	•					•						
Arch.7	Is access control provided through main entrance points for employees and visitors? (lobby receptionist, sign-in, staff escorts, issue of visitor badges, checking forms of personal identification, electronic access control systems)	•	•		•			•						
Arch.8	Is access to private and public space or restricted area space clearly defined through the design of the space, signage, use of electronic security devices, etc.?	•	•		•			•						
Arch.9	Is access to elevators distinguished as to those that are designated only for employees and visitors?	•	•		•			•						
Arch.10	Do public and employee entrances include space for possible future installation of access control and screening equipment?	•	•		•			•						
Arch.11	Do foyers have reinforced concrete walls and offset interior and exterior doors from each other?	•	•										•	
Arch.12	Do doors and walls along the line of security screening meet requirements of UL752 "Standard for Safety: Bullet-Resisting Equipment"?	•	•	•										

Source: Federal Emergency Management Agency (FEMA)

Table B: Architectural (continued)

Item	Vulnerability Question	Vulnerability estimate	Detailed assessment	Visual inspection	Document review	Org/Mgmt procedure	Moving vehicle	Stationary vehicle	Covert entry	Mail	Supplies	Blast effects	Airborne (contamination)	Waterborne (contamination)
		Characterization					**Terrorist Tactics**							
		Type		Collection			Delivery Methods					Mechanisms		
Arch.13	Do circulation routes have unobstructed views of people approaching controlled access points?	•		•					•					
Arch.14	Is roof access limited to authorized personnel by means of locking mechanisms?		•	•		•			•					
Arch.15a	Are critical assets (people, activities, building systems and components) located close to any main entrance, vehicle circulation, parking, maintenance area, loading dock, or interior parking?	•		•	•	•	•	•	•			•		
Arch.15b	Are the critical building systems and components hardened?		•	•	•							•		
Arch.16	Are high-value or critical assets located as far into the interior of the building as possible and separated from the public areas of the building?	•		•	•	•						•		
Arch.17	Is high visitor activity away from critical assets?	•				•		•						
Arch.18a	Are critical assets located in spaces that are occupied 24 hours per day?	•				•								
Arch.18b	Are assets located in areas where they are visible to more than one person?	•				•								
Arch.19	Are loading docks and receiving and shipping areas separated in any direction from utility rooms, utility mains, and service entrances including electrical, telephone/data, fire detection/alarm systems, fire suppression water mains, cooling and heating mains, etc.?		•	•							•	•	•	•
Arch.20a	Are mailrooms located away from facility main entrances, areas containing critical services, utilities, distribution systems, and important assets?		•	•						•				
Arch.20b	Is the mailroom located near the loading dock?		•	•						•	•	•	•	•
Arch.21	Does the mailroom have adequate space available for equipment to examine incoming packages and for an explosive disposal container?		•	•							•	•	•	•
Arch.22	Are areas of refuge identified, with special consideration given to egress?	•	•	•	•									

Source: Federal Emergency Management Agency (FEMA)

Table B: Architectural (continued)

		Characterization				Terrorist Tactics						
		Type	Collection			Delivery Methods				Mechanisms		
Item	Vulnerability Question	Vulnerability estimate	Detailed assessment	Visual inspection	Document review	Org/Mgmt procedure	Moving vehicle	Stationary vehicle	Covert entry	Mail	Supplies	Blast effects	Airborne (contamination)	Waterborne (contamination)
Arch.23a	Are stairwells required for emergency egress located as remotely as possible from high-risk areas where blast events might occur?	●	●									●		
Arch.23b	Are stairways maintained with positive pressure or are there other smoke control systems?	●	●	●									●	
Arch.24	Are enclosures for emergency egress hardened to limit the extent of debris that might otherwise impede safe passage and reduce the flow of evacuees?	●	●									●		
Arch.25	Do interior barriers differentiate level of security within a facility?	●		●	●	●								
Arch.26	Are emergency systems located away from high-risk areas?	●		●	●	●						●		
Arch.27a	Is interior glazing near high-threat areas minimized?	●	●									●		
Arch.27b	Is interior glazing in other areas shatter resistant?	●		●								●		

Source: Federal Emergency Management Agency (FEMA)

Table C: Structural Systems

Item	Vulnerability Question	Vulnerability estimate	Detailed assessment	Visual inspection	Document review	Org/Mgmt procedure	Moving vehicle	Stationary vehicle	Covert entry	Mail	Supplies	Blast effects	Airborne (contamination)	Waterborne (contamination)
StrucSys.1a	What type of construction?	●		●	●							●		
StrucSys.1b	What type of concrete & reinforcing steel?		●		●							●		
StrucSys.1c	What type of steel?		●		●							●		
StrucSys.1d	What type of foundation?		●		●							●		
StrucSys.2a	Do the reinforced concrete structures contain symmetric steel reinforcement (positive and negative faces) in all floor slabs, roof slabs, walls, beams and girders that may be subjected to rebound, uplift and suction pressures?		●		●							●		
StrucSys.2b	Do the lap splices fully develop the capacity of the reinforcement?		●		●							●		
StrucSys.2c	Are lap splices and other discontinuities staggered?		●		●							●		
StrucSys.2d	Do the connections possess ductile details?		●		●							●		
StrucSys.2e	Is special shear reinforcement, including ties and stirrups, available to allow large post-elastic behavior?		●		●							●		
StrucSys.3a	Are the steel frame connections moment connections?		●	●	●							●		
StrucSys.3b	Is the column spacing minimized so that reasonably sized members will resist the design loads and increase the redundancy of the system?	●		●	●							●		
StrucSys.3c	What are the floor-to-floor heights?	●		●	●							●		
StrucSys.4	Are critical elements vulnerable to failure?		●		●							●		
StrucSys.5	Will the structure suffer an unacceptable level of damage resulting from the postulated threat (blast loading or weapon impact)?		●		●							●		
StrucSys.6a	Is the structure vulnerable to progressive collapse?	●		●	●							●		
StrucSys.6b	Is the facility capable of sustaining the removal of a column for one floor above grade at the building perimeter without progressive collapse?		●		●							●		

Source: Federal Emergency Management Agency (FEMA)

Table C: Structural Systems (continued)

Item	Vulnerability Question	Vulnerability estimate	Detailed assessment	Visual inspection	Document review	Org/Mgmt procedure	Moving vehicle
		Characterization					
		Type		Collection			
StrucSys.6c	In the event of an internal explosion in an uncontrolled public ground floor area does the design prevent progressive collapse due to the loss of one primary column?	●		●			
StrucSys.6d	Do architectural or structural features provide a minimum 6-inch standoff to the internal columns?	●	●	●			
StrucSys.6e	Are the columns in the unscreened internal spaces designed for an unbraced length equal to two floors, or three floors where there are two levels of parking?	●		●			
StrucSys.7	Are there adequate redundant load paths in the structure?	●			●		
StrucSys.8	Are there transfer girders supported by columns within unscreened public spaces or at the exterior of the building?	●	●				
StrucSys.9	Will the loading dock design limit damage to adjacent areas and vent explosive force to the exterior of the building?	●	●	●			
StrucSys.10	Are mailrooms, where packages are received and opened for inspection, and unscreened retail spaces designed to mitigate the effects of a blast on primary vertical or lateral bracing members?	●			●	●	

Source: Federal Emergency Management Agency (FEMA)

Table D:Building Envelope

Item	Vulnerability Question	Vulnerability estimate	Detailed assessment	Visual inspection	Document review	Org/Mgmt procedure	Moving vehicle	Stationary vehicle	Covert entry	Mail	Supplies	Blast effects	Airborne (contamination)	Waterborne (contamination)
		Characterization					Terrorist Tactics							
		Type		Collection			Delivery Methods					Mechanisms		
BldgEnv.1	What is the designed or estimated protection level of the exterior walls against the postulated explosive threat?	●		●								●		
BldgEnv.2a	Is there less than 40 % fenestration openings per structural bay?	●	●									●		
BldgEnv.2b	Is the window system design on the exterior façade balanced to mitigate the hazardous effects of flying glazing following an explosive event? (glazing, frames, anchorage to supporting walls, etc.)	●	●	●								●		
BldgEnv.2c	Do the glazing systems with a ½-inch bite contain an application of structural silicone?	●	●	●								●		
BldgEnv.2d	Is the glazing laminated or is it protected with an anti-shatter film?	●	●	●								●		
BldgEnv.2e	If an anti-shatter film is used, is it a minimum of a 7-mil thick film, or specially manufactured 4-mil thick film?	●	●	●								●		
BldgEnv.3a	Do the walls, anchorage, and window framing fully develop the capacity of the glazing material selected?	●		●								●		
BldgEnv.3b	Are the walls capable of withstanding the dynamic reactions from the windows?	●		●								●		
BldgEnv.3c	Will the anchorage remain attached to the walls of the facility during an explosive event without failure?	●		●								●		
BldgEnv.3d	Is the façade connected to back-up block or to the structural frame?	●	●	●								●		
BldgEnv.3e	Are non-bearing masonry walls reinforced?	●	●	●								●		
BldgEnv.4a	Does the facility contain ballistic glazing?	●	●	●								●		
BldgEnv.4b	Does the ballistic glazing meet the requirements of UL 752 Bullet-Resistant Glazing?	●	●	●										
BldgEnv.4c	Does the facility contain security-glazing?	●	●	●					●					
BldgEnv.4d	Does the security-glazing meet the requirements of ASTM F1233 or UL 972, Burglary Resistant Glazing Material?	●	●	●					●					
BldgEnv.4e	Do the window assemblies containing forced entry resistant glazing (excluding the glazing) meet the requirements of ASTM F 588?	●	●	●					●					
BldgEnv.5	Do non-window openings, such as mechanical vents and exposed plenums, provide the same level of protection required for the exterior wall?	●	●	●									●	●

Source: Federal Emergency Management Agency (FEMA)

Table E: Utility Systems

Item	Vulnerability Question	Vulnerability estimate	Detailed assessment	Visual inspection	Document review	Org/Mgmt procedure	Moving vehicle	Stationary vehicle	Covert entry	Mail	Supplies	Blast effects	Airborne (contamination)	Waterborne (contamination)
UtilSys.1a	What is the source of domestic water? (utility, municipal, wells, lake, river, storage tank)	•		•	•									•
UtilSys.1b	Is there a secure alternate drinking water supply?		•	•	•	•								•
UtilSys.2	Are there multiple entry points for the water supply?		•	•	•									•
UtilSys.3	Is the incoming water supply in a secure location?		•	•	•	•								•
UtilSys.4a	Does the facility have storage capacity for domestic water?		•	•	•									•
UtilSys.4b	How many gallons and how long will it allow operations to continue?	•		•	•	•								•
UtilSys.5a	What is the source of water for the fire suppression system? (local utility company lines, storage tanks with utility company backup, lake, or river)	•		•	•									
UtilSys.5b	Are there alternate water supplies for fire suppression?	•		•	•	•								
UtilSys.6	Is the fire suppression system adequate, code-compliant, and protected (secure location)?		•	•	•									
UtilSys.7a	Do the sprinkler/standpipe interior controls (risers) have fire- and blast-resistant separation?		•	•								•		
UtilSys.7b	Are the sprinkler and standpipe connections adequate and redundant?	•		•	•									
UtilSys.7c	Are there fire hydrant and water supply connections near the sprinkler/standpipe connections?		•	•										
UtilSys.8a	Are there redundant fire water pumps (e.g., one electric, one diesel)?		•	•		•								
UtilSys.8b	Are the pumps located apart from each other?		•	•										
UtilSys.9a	Are sewer systems accessible?		•	•	•	•								
UtilSys.9b	Are they protected or secured?		•	•	•									
UtilSys.10	What fuel supplies do the facility rely upon for critical operation?	•		•	•	•								
UtilSys.11a	How much fuel is stored on the site or at the facility and how long can this quantity support critical operations?	•	•			•								

Source: Federal Emergency Management Agency (FEMA)

Table E: Utility Systems (continued)

| | | Characterization | | | | | Terrorist Tactics | | | | | | | |
| | | Type | | Collection | | | Delivery Methods | | | | | Mechanisms | | |
Item	Vulnerability Question	Vulnerability estimate	Detailed assessment	Visual inspection	Document review	Org/Mgmt procedure	Moving vehicle	Stationary vehicle	Covert entry	Mail	Supplies	Blast effects	Airborne (contamination)	Waterborne (contamination)
UtilSys.11b	How is it (fuel) stored?		•	•										
UtilSys.11c	How is it (fuel) secured?		•	•		•								
UtilSys.12a	Where is the fuel supply obtained?	•				•								
UtilSys.12b	How is it (fuel) delivered?		•			•								
UtilSys.13a	Are there alternate sources of fuel?	•				•								
UtilSys.13a	Are there alternate sources of fuel?	•				•								
UtilSys.13b	Can alternate fuels be used?	•			•	•								
UtilSys.14	What is the normal source of electrical service for the facility?	•		•	•									
UtilSys.15a	Is there a redundant electrical service source?		•		•	•								
UtilSys.15b	Can the facilities be fed from more than one utility substation?		•			•								
UtilSys.16	How may service entry points does the facility have for electricity?		•	•	•									
UtilSys.17	Is the incoming electric service to the building secure?		•	•		•								
UtilSys.18a	What provisions for emergency power exist? What systems receive emergency power and have capacity requirements been tested?	•		•	•	•								
UtilSys.18b	Is the emergency power co-located with the commercial electric service?		•	•										
UtilSys.18c	Is there an exterior connection for emergency power?		•	•										
UtilSys.19	By what means does the main telephone and data communications interface the facility?	•		•	•	•								
UtilSys.20	Are there multiple or redundant locations for the telephone and communication service?		•	•	•	•								
UtilSys.21a	Does the fire alarm system require communication with external sources?		•		•	•								
UtilSys.21b	By what method is the alarm signal sent to the responding agency: telephone, radio, etc?		•		•	•								
UtilSys.21c	Is there an intermediary alarm monitoring center?		•		•	•								
UtilSys.22	Are utility lifelines aboveground, underground, or direct buried?		•	•	•									

Source: Federal Emergency Management Agency (FEMA)

Table F: Mechanical Systems (Including Chemical, Biological, and Radiological Systems)

Item	Vulnerability Question	Vulnerability estimate	Detailed assessment	Visual inspection	Document review	Org/Mgmt procedure	Moving vehicle	Stationary vehicle	Covert entry	Mail	Supplies	Blast effects	Airborne (contamination)	Waterborne (contamination)
		Type		**Collection**			**Delivery Methods**					**Mechanisms**		
MechSys.1a	Where are the air intakes and exhaust louvers for the building? (low, high, or midpoint of the building structure)	•		•	•								•	
MechSys.1b	Are the intakes accessible to the public?		•	•		•							•	
MechSys.2	Are there multiple air intake locations?	•		•	•								•	
MechSys.3a	What are the types of air filtration? Include the efficiency and number of filter modules for each of the main air handling systems.	•			•								•	
MechSys.3b	Is there any collective protection for chemical, biological, and radiological contamination designed into the facility?	•	•	•	•								•	
MechSys.4	Is there space for larger filter assemblies on critical air handling systems?	•	•	•									•	
MechSys.5	Are there provisions for air monitors or sensors for chemical or biological agents?	•	•	•									•	
MechSys.6	By what method are air intakes closed when not operational?	•				•							•	
MechSys.7a	How are air handling systems zoned?	•		•	•								•	
MechSys.7b	What areas and functions do each of the primary air handling systems serve?	•			•								•	
MechSys.8	Are there large central air handling units or are there multiple units serving separate zones?	•			•								•	
MechSys.9a	Are there any redundancies in the air handling system?	•			•	•							•	
MechSys.9b	Can critical areas be served from other units if a major system is disabled?	•			•	•							•	
MechSys.10	Is the air supply to critical areas compartmentalized?	•			•	•							•	
MechSys.11	Are supply and exhaust air systems for critical areas secure?	•			•	•							•	
MechSys.12a	What is the method of temperature and humidity control?	•	•	•	•								•	
MechSys.12b	Is it (temp. control) localized or centralized?	•	•	•	•									
MechSys.13a	Where are the building automation control centers and cabinets located?	•	•	•										

Source: Federal Emergency Management Agency (FEMA)

Table F: Mechanical Systems (Including Chemical, Biological, and Radiological Systems) (continued)

Item	Vulnerability Question	Vulnerability estimate	Detailed assessment	Visual inspection	Document review	Org/Mgmt procedure	Moving vehicle	Stationary vehicle	Covert entry	Mail	Supplies	Blast effects	Airborne (contamination)	Waterborne (contamination)
		Characterization					Terrorist Tactics							
		Type		Collection			Delivery Methods					Mechanisms		
MechSys.13b	Are they in secure areas?		●	●	●	●								
MechSys.13c	How is the control wiring routed?		●	●	●									
MechSys.14	Does the control of air handling systems support plans for sheltering in place?		●		●	●								
MechSys.15	Where is roof-mounted equipment located on the roof?(near perimeter, at center of roof)	●		●										
MechSys.16	Are fire dampers installed at all fire barriers?		●	●	●									
MechSys.17	Do fire walls and fire doors maintain their integrity?		●	●	●									
MechSys.18	Do elevators have recall capability and elevator emergency message capability?		●	●	●	●								

Table G: Plumbing and Gas Systems

Item	Vulnerability Question	Vulnerability estimate	Detailed assessment	Visual inspection	Document review	Org/Mgmt procedure	Moving vehicle	Stationary vehicle	Covert entry	Mail	Supplies	Blast effects	Airborne (contamination)	Waterborne (contamination)
		Characterization					Terrorist Tactics							
		Type		Collection			Delivery Methods					Mechanisms		
PlumbGas.1	What is the method of water distribution?	●		●										●
PlumbGas.2	What is the method of gas distribution?, (heating, cooking, medical, process)	●		●										
PlumbGas.3	Is there redundancy to the main piping distribution?		●	●										
PlumbGas.4a	What is the method of heating domestic water?	●		●	●	●								
PlumbGas.4b	What fuel(s) is used?		●	●	●	●								
PlumbGas.5a	Where are gas storage tanks located? (heating, cooking, medical, process)		●	●										
PlumbGas.5b	How are they (gas tanks) piped to the distribution system?(above or below ground)		●	●	●									
PlumbGas.6	Are there reserve supplies of critical gases?	●			●	●								

Source: Federal Emergency Management Agency (FEMA)

Table H: Electrical Systems

Item	Vulnerability Question	Vulnerability estimate	Detailed assessment	Visual inspection	Document review	Org/Mgmt procedure	Moving vehicle	Stationary vehicle	Covert entry	Mail	Supplies	Blast effects	Airborne (contamination)	Waterborne (contamination)
		Type					Delivery Methods						Mechanisms	
		Characterization					Terrorist Tactics							
ElectSys.1a	Are there any transformers or switchgears located outside the building or accessible from the building exterior?	●		●										
ElectSys.1b	Are they (transformers or switchgears) vulnerable to public access?	●		●		●								
ElectSys.1c	Are they (transformers or switchgears) secured?		●	●		●								
ElectSys.2	What is the extent of the external facility lighting in utility and service areas and at normal entryways used by the building occupants?		●	●										
ElectSys.3	How are the electrical rooms secured and where are they located relative to other higher risk areas, starting with the main electrical distribution room at the service entrance?		●	●	●	●								
ElectSys.4a	Are critical electrical systems co-located with other building systems?		●	●										
ElectSys.4b	Are critical electrical systems located in areas outside of secured electrical areas?	●		●	●	●								
ElectSys.4c	Is security system wiring located separately from electrical and other service systems?		●	●	●									
ElectSys.5	How are electrical distribution panels serving branch circuits secured or are they in secure locations?		●	●	●	●								
ElectSys.6a	Does emergency backup power exist for all areas within the facility or for critical areas only?	●		●	●									
ElectSys.6b	How is the emergency power distributed?		●		●	●								
ElectSys.6c	Is the emergency power system independent from the normal electrical service, particularly in critical areas?		●	●										
ElectSys.7a	How is the primary electrical system wiring distributed?	●			●	●								
ElectSys.7b	Is it co-located with other major utilities?		●	●	●									
ElectSys.7c	Is there redundancy of distribution to critical areas?		●	●	●	●								

Source: Federal Emergency Management Agency (FEMA)

Table I: Fire Alarm Systems

| Item | Vulnerability Question | Characterization | | | | | Terrorist Tactics | | | | | | | |
| | | Type | | Collection | | | Delivery Methods | | | | | Mechanisms | | |
		Vulnerability estimate	Detailed assessment	Visual inspection	Document review	Org/Mgmt procedure	Moving vehicle	Stationary vehicle	Covert entry	Mail	Supplies	Blast effects	Airborne (contamination)	Waterborne (contamination)
FireAlarm.1a	Is the facility fire alarm system centralized or localized?		●	●	●									
FireAlarm.1b	How are alarms annunciated, both locally and centrally?		●	●	●	●								
FireAlarm.1c	Are critical documents and control systems located in a secure yet accessible location?		●	●		●								
FireAlarm.2a	Where are the fire alarm panels located?		●	●										
FireAlarm.2b	Are they allow access to unauthorized personnel?		●	●		●								
FireAlarm.3a	Is the fire alarm system stand-alone or integrated with other functions such as security and environmental or building management systems?	●			●	●								
FireAlarm.3b	What is the interface?			●	●	●								
FireAlarm.4	Do key fire alarm system components have fire- and blast-resistant separation?		●	●	●							●		
FireAlarm.5	Is there redundant off-premises fire alarm reporting?	●			●	●								

Source: Federal Emergency Management Agency (FEMA)

Table J: Communications and IT Systems

Item	Vulnerability Question	Vulnerability estimate	Detailed assessment	Visual inspection	Document review	Org/Mgmt procedure	Moving vehicle	Stationary vehicle	Covert entry	Mail	Supplies	Blast effects	Airborne (contamination)	Waterborne (contamination)
		Characterization — Type		Collection			Terrorist Tactics — Delivery Methods					Mechanisms		
CommIT.1a	Where is the main telephone distribution room and where is it in relation to higher risk areas?	●		●	●	●								
CommIT.1b	Is the main telephone distribution room secure?		●	●		●								
CommIT.2a	Does the telephone system have an UPS (uninterruptible power supply)?		●		●	●								
CommIT.2b	What is its (ups) type, power rating, operational duration under load, and location? (battery, on-line, filtered)		●		●									
CommIT.3a	Where are communication systems wiring closets located? (voice, data, signal, alarm)	●		●	●									
CommIT.3b	Are they (communication closets) co-located with other utilities?		●	●	●									
CommIT.3c	Are they (communication closets) in secure areas?		●	●	●	●								
CommIT.4	How is communications system wiring distributed? (secure chases and risers, accessible public areas)		●	●	●									
CommIT.5	Are there redundant communications systems available?		●		●	●								
CommIT.6a	Where are the main distribution facility, data centers, routers, firewalls, and servers located?		●		●									
CommIT.6b	Where are the secondary and/or intermediate (IT) distribution facilities?		●		●									
CommIT.7	What type and where are the WAN (wide area network) connections?		●	●	●									
CommIT.8a	What type, power rating, and location of the UPS (uninterruptible power supply)? (battery, on-line, filtered)		●	●	●									
CommIT.8b	Are the UPS also connected to emergency power?		●	●	●									
CommIT.9	What type of LAN (local area network) cabling and physical topology is used? (Category(Cat) 5, Gigabit Ethernet, Ethernet, Token Ring)		●	●	●									
CommIT.10	For installed radio/wireless systems, what are their types and where are they located? (RF (radio frequency), HF (high frequency), VHF (very high frequency), MW (medium wave)		●	●	●									

Source: Federal Emergency Management Agency (FEMA)

Table J: Communications and IT Systems (continued)

Item	Vulnerability Question	Vulnerability estimate	Detailed assessment	Visual inspection	Document review	Org/Mgmt procedure	Moving vehicle	Stationary vehicle	Covert entry	Mail	Supplies	Blast effects	Airborne (contamination)	Waterborne (contamination)
		Characterization					**Terrorist Tactics**							
		Type	Collection				Delivery Methods					Mechanisms		
CommIT.11	Do the IT (Information Technology - computer) systems meet requirements of confidentiality, integrity, and availability?	●			●	●								
CommIT.12	Where is the disaster recovery/mirroring site?	●			●	●								
CommIT.13	Where is the back-up tape/file storage site and what is the type of safe environment? (safe, vault, underground)Is there redundant refrigeration in the (backup IT storage) site?	●			●	●								
CommIT.14	Are there any SATCOM (satellite communications) links? (location, power, UPS, emergency power, spare capacity/capability)	●	●	●										
CommIT.15a	Is there a mass notification system that reaches all building occupants? (public address, pager, cell phone, computer override, etc.)	●				●								
CommIT.15b	Will one or more of these systems be operational under hazard conditions? (UPS, emergency power)	●			●	●								
CommIT.16a	Do control centers and their designated alternate locations have equivalent or reduced capability for voice, data, mass notification, etc.? (emergency operations, security, fire alarms, building automation)	●			●	●								
CommIT.16b	Do the alternate locations also have access to backup systems, including emergency power?	●				●								

Source: Federal Emergency Management Agency (FEMA)

Table K: Equipment Operations and Maintenance

		Characterization		Collection			Delivery Methods					Mechanisms		
Item	Vulnerability Question	Vulnerability estimate	Detailed assessment	Visual inspection	Document review	Org/Mgmt procedure	Moving vehicle	Stationary vehicle	Covert entry	Mail	Supplies	Blast effects	Airborne (contamination)	Waterborne (contamination)
EquipOM.1a	Are there composite drawings indicating location and capacities of major systems and are they current? (electrical, mechanical, and fire protection; and date of last update)	●		●										
EquipOM.1b	Do updated O&M (operation and maintenance) manuals exist?	●		●										
EquipOM.2a	Have critical air systems been rebalanced?	●			●	●								
EquipOM.2b	If so, when and how often?	●				●								
EquipOM.3	Is air pressurization monitored regularly?	●				●								
EquipOM.4	Does the facility have a policy or procedure for periodic recommissioning of major Mechanical/Electrical/Plumbing systems?	●				●								
EquipOM.5	Is there an adequate operations and maintenance program including training of facilities management staff?	●				●								
EquipOM.6	What maintenance and service agreements exist for M/E/P systems?	●			●									
EquipOM.7	Are backup power systems periodically tested under load?	●				●								
EquipOM.8	Is stairway and exit sign lighting operational?	●	●											

Source: Federal Emergency Management Agency (FEMA)

Table L: Security Systems

Item	Vulnerability Question	Characterization — Type: Vulnerability estimate	Characterization — Type: Detailed assessment	Characterization — Collection: Visual inspection	Characterization — Collection: Document review	Characterization — Collection: Org/Mgmt procedure	Terrorist Tactics — Delivery Methods: Moving vehicle	Terrorist Tactics — Delivery Methods: Stationary vehicle	Terrorist Tactics — Delivery Methods: Covert entry	Terrorist Tactics — Delivery Methods: Mail	Terrorist Tactics — Delivery Methods: Supplies	Terrorist Tactics — Mechanisms: Blast effects	Terrorist Tactics — Mechanisms: Airborne (contamination)	Terrorist Tactics — Mechanisms: Waterborne (contamination)
SecPerim.1a	Are black/white or color CCTV (closed circuit television) cameras used?	●	●						●	●				
SecPerim.1b	Are they monitored and recorded 24 hours/7 days a week? By whom?	●			●	●			●	●				
SecPerim.1c	Are they analog or digital by design?	●		●					●	●				
SecPerim.1d	What are the number of fixed, wireless and pan-tilt-zoom cameras used?	●		●	●				●	●				
SecPerim.1e	Who are the manufacturers of the CCTV cameras?	●		●	●				●	●				
SecPerim.1f	What is the age of the CCTV cameras in use?	●			●	●			●	●				
SecPerim.2a	Are the cameras programmed to respond automatically to perimeter building alarm events?	●			●	●			●	●				
SecPerim.2b	Do they have built-in video motion capabilities?	●		●					●	●				
SecPerim.3	What type of camera housings are used and are they environmental in design to protect against exposure to heat and cold weather elements?	●		●	●				●	●				
SecPerim.4	Are panic/duress alarm buttons or sensors used, where are they located and are they hardwired or portable?	●	●						●	●				
SecPerim.5	Are intercom call boxes used in parking areas or along the building perimeter?	●	●		●	●			●	●				
SecPerim.6	What is the transmission media used to transmit camera video signals: fiber, wire line, telephone wire, coaxial, wireless?	●		●	●				●	●				
SecPerim.7	Who monitors the CCTV system?	●			●	●			●	●				
SecPerim.8a	What is the quality of video images both during the day and hours of darkness?	●	●		●	●			●	●				
SecPerim.8b	Are infrared camera illuminators used?	●	●		●	●			●	●				
SecPerim.9	Are the perimeter cameras supported by an uninterruptible power supply, battery, or building emergency power?	●		●					●	●				
SecPerim.10	What type of exterior IDS sensors are used: electromagnetic, fiber optic, active infrared, bistatic microwave, seismic, photoelectric, ground, fence, glass break (vibration/shock), single, double and roll-up door magnetic contacts or switches.	●		●	●				●	●				

Source: Federal Emergency Management Agency (FEMA)

Table L: Security Systems (continued)

Item	Vulnerability Question	Vulnerability estimate	Detailed assessment	Visual inspection	Document review	Org/Mgmt procedure	Moving vehicle	Stationary vehicle	Covert entry	Mail	Supplies	Blast effects	Airborne (contamination)	Waterborne (contamination)
				Characterization					Terrorist Tactics					
			Type	Collection			Delivery Methods					Mechanisms		
SecPerim.11	Is a global positioning satellite system (GPS) used to monitor vehicles and asset movements?	•		•	•	•	•							
SecInter.12a	Are black/white or color CCTV (closed circuit television) cameras used?	•	•						•					
SecInter.12b	Are they monitored and recorded 24 hours/7 days a week? By whom?	•			•				•					
SecInter.12c	Are they analog or digital by design?	•		•					•					
SecInter.12d	What are the number of fixed, wireless and pan-tilt-zoom cameras used?	•	•	•					•					
SecInter.12e	Who are the manufacturers of the CCTV cameras?	•	•	•					•					
SecInter.12f	What is the age of the CCTV cameras in use?	•		•	•				•					
SecInter.13a	Are the cameras programmed to respond automatically to perimeter building alarm events?	•			•				•					
SecInter.13b	Do they have built-in video motion capabilities?	•		•					•					
SecInter.14	What type of camera housings are used and are they designed to protect against exposure or tampering?	•	•	•					•					
SecInter.15	Are the camera lenses used of the proper specifications, especially distance viewing and clarity?	•		•	•				•					
SecInter.16	What is the transmission media used to transmit camera video signals: fiber, wire line, telephone wire, coaxial, wireless?	•	•	•					•					
SecInter.17	Is the quality in interior camera video images of good visual and recording quality?	•	•		•				•					
SecInter.18	Are the interior cameras supported by an uninterruptible power supply source, battery, or building emergency power?	•	•						•					
SecInter.19	What are the first costs and maintenance costs associated with the interior cameras?	•		•	•				•					
SecInter.20a	What type of security access control system is used?	•			•				•					
SecInter.20b	Are these same devices used for physical security also used (integrated) with providing access control to security computer networks (e.g. in place of or in combination with user ID and system passwords)?	•		•	•				•					

Source: Federal Emergency Management Agency (FEMA)

Table L: Security Systems (continued)

| | | Characterization | | | | | Terrorist Tactics | | | | | | | |
| | | Type | | Collection | | | Delivery Methods | | | | | Mechanisms | | |
Item	Vulnerability Question	Vulnerability estimate	Detailed assessment	Visual inspection	Document review	Org/Mgmt procedure	Moving vehicle	Stationary vehicle	Covert entry	Mail	Supplies	Blast effects	Airborne (contamination)	Waterborne (contamination)
SecInter.23b	How old are the systems and what are the related first and maintenance service costs?	•			•	•			•					
SecInter.24	Are there panic/duress alarm sensors used, where are they located and are they hardwired or portable?	•	•	•					•					
SecInter.25	Are intercom call-boxes or building intercom system used throughout the facility?	•	•	•	•				•					
SecInter.26	Are magnetometers (metal detectors) and x-ray equipment used and at what locations within the facility?	•	•						•					
SecInter.27	What type of interior IDS sensors are used: electromagnetic, fiber optic, active infrared-motion detector, photoelectric, glass break (vibration/shock), single, double and roll-up door magnetic contacts or switches?	•	•	•					•					
SecInter.28	Are mechanical, electrical, medical gas, power supply, radiological material storage, voice/data telecommunication system nodes, security system panels, elevator and critical system panels, and other sensitive rooms continuously locked, under electronic security CCTV camera and intrusion alarm systems surveillance?	•				•			•					
SecInter.29	What types of locking hardware are used throughout the facility? Are manual and electromagnetic cipher, keypad, pushbutton, panic bar, door strikes and related hardware and software used?	•	•						•					
SecInter.30	Are any potentially hazardous chemicals, combustible or toxic materials stored on-site in non-secure and non-monitored areas?	•				•								
SecInter.31	What security controls are in place to handle the processing of mail and protect against potential biological, explosive or other threatening exposures?	•				•								
SecInter.32a	Is there a designated security control room and console in place to monitor security, fire alarm and possibly other building systems?	•	•						•					
SecInter.32b	Is there a backup control center designated and equipped?	•	•	•					•					

Source: Federal Emergency Management Agency (FEMA)

Table L: Security Systems (continued)

Item	Vulnerability Question	Vulnerability estimate	Detailed assessment	Visual inspection	Document review	Org/Mgmt procedure	Moving vehicle	Stationary vehicle	Covert entry	Mail	Supplies	Blast effects	Airborne (contamination)	Waterborne (contamination)
SecInter.32c	Is there off-site 24-hour monitoring of intrusion detection systems?	•			•				•					
SecInter.33	Is the security console and control room adequate in size, provide room for expansion, have adequate environment controls (e.g. a/c, lighting, heating, air circulation, backup power, etc,) and is ergonomically designed?	•	•											
SecInter.34	Is the location of the security room in a secure area with limited, controlled and restricted access controls in place?	•	•		•									
SecInter.35a	What are the means by which facility and security personnel can communicate with one another: portable radio, pager, cell phone, personal data assistants (PDA's), etc)?	•			•									
SecInter.35b	What problems have been experienced with these and other electronic security systems?	•			•									
SecInter.36	Is there a computerized security incident reporting system used to prepare reports and track security incident trends and patterns?	•			•									
SecInter.37	Does the present security force have access to use a computerized guard tour system?	•			•									
SecInter.38a	Are vaults or safes in the facility?	•	•											
SecInter.38b	Where are they located?	•	•											
SecDocs.39	Are security system as-built drawings been generated and ready for review?	•			•									
SecDocs.40	Have security system design and drawing standards been developed?	•			•									
SecDocs.41	Are security equipment selection criteria defined?	•			•									
SecDocs.42	What contingency plans have been developed or are in place to deal with security control center redundancy and backup operations?	•			•									
SecDocs.43	Have security system construction specification documents been prepared and standardized?	•			•									
SecDocs.44	Are all security system documents to include as-built drawings current?	•			•									

Source: Federal Emergency Management Agency (FEMA)

Table L: Security Systems (continued)

Item	Vulnerability Question	Characterization — Type — Vulnerability estimate	Detailed assessment	Collection — Visual inspection	Document review	Org/Mgmt procedure	Terrorist Tactics — Delivery Methods — Moving vehicle	Stationary vehicle	Covert entry	Mail	Supplies	Mechanisms — Blast effects	Airborne (contamination)	Waterborne (contamination)
SecDocs.45	Have qualifications been determined in using security consultants, system designers and engineers, installation vendors, and contractors?	●				●								
SecDocs.46	Are security systems decentralized, centralized, integrated, and operate over existing IT network or standalone method of operation?	●			●	●								
SecDocs.47	What security systems manuals are available?	●		●										
SecDocs.48	What maintenance or service agreements exist for security systems?	●		●										

Source: Federal Emergency Management Agency (FEMA)

Table M: Security Master Plan

Item	Vulnerability Question	Vulnerability estimate	Detailed assessment	Visual inspection	Document review	Org/Mgmt procedure	Moving vehicle	Stationary vehicle	Covert entry	Mail	Supplies	Blast effects	Airborne (contamination)	Waterborne (contamination)
		Characterization					Terrorist Tactics							
		Type		Collection			Delivery Methods					Mechanisms		
SecPlan.1a	Does a written security plan exist for this facility?		●		●									
SecPlan.1b	When was the initial security plan written and last revised?		●		●									
SecPlan.1c	Who is responsible for preparing and reviewing the security plan?		●			●								
SecPlan.2	Has the security plan been communicated and disseminated to key management personnel and departments?		●			●								
SecPlan.3	Has the security plan been benchmarked or compared against related organizations and operational entities?		●			●								
SecPlan.4	Has the security plan ever been tested and evaluated from a cost-benefit and operational efficiency and effectiveness perspective?		●			●								
SecPlan.5	Does it define mission, vision, short-long term security program goals and objectives?		●			●								
SecPlan.6	Are threats, vulnerabilities, risks adequately defined and security countermeasures addressed and prioritized relevant to their criticality and probability of occurrence?		●			●								
SecPlan.7	Has a security implementation schedule been established to address recommended security solutions?		●			●								
SecPlan.8	Have security operating and capital budgets been addressed, approved and established to support the plan?		●		●	●								
SecPlan.9	What regulatory or industry guidelines/standards were followed in the preparation of the security plan?		●		●	●								
SecPlan.10	Does the security plan address existing security conditions from an administrative, operational, managerial and technical security systems perspective?		●		●	●								
SecPlan.11	Does the security plan address the protection of people, property, assets, and information?		●		●	●								
SecPlan.12	Does the security plan address the following major components: access control, surveillance, response, building hardening and protection against biological, chemical, radiological and cyber-network attacks?		●		●	●								

Source: Federal Emergency Management Agency (FEMA)

Table M: Security Master Plan (continued)

Item	Vulnerability Question	Vulnerability estimate	Detailed assessment	Visual inspection	Document review	Org/Mgmt procedure	Moving vehicle	Stationary vehicle	Covert entry	Mail	Supplies	Blast effects	Airborne (contamination)	Waterborne (contamination)
		Characterization					Terrorist Tactics							
		Type		Collection			Delivery Methods					Mechanisms		
SecPlan.13	Has the level of risk been identified and communicated in the security plan through the performance of a physical security assessment?	●			●	●								
SecPlan.14a	When was the last security assessment performed?	●			●	●								
SecPlan.14b	Who performed the security risk assessment?	●			●	●								
SecPlan.15a	Were the following areas of security analysis addressed in the security master plan: Asset Analysis: Does the security plan identify and prioritize the assets to be protected in accordance to their location, control, current value, and replacement value?	●			●									
SecPlan.15b	Threat Analysis: Does the security plan address potential threats; causes of potential harm in the form of death, injury, destruction, disclosure, interruption of operations, or denial of services? (possible criminal acts (documented and review of police/security incident reports) associated with forced entry, bombs, ballistic assault, biochemical and related terrorist tactics, attacks against utility systems infrastructure and buildings)	●			●									
SecPlan.15c	Vulnerability Analysis: Does the security plan address other areas and anything else associated with a facility and it's operations that can be taken advantage of to carry out a threat? (architectural design and construction of new and existing facilities, technological support systems (e.g. heating, air conditioning, power, lighting and security systems, etc.) and operational procedures, policies and controls)	●			●									
SecPlan.15d	Risk Analysis: Does the security plan address the findings from the asset, threat, and vulnerability analyses to develop, recommend and consider implementation of appropriate security countermeasures?	●			●									

Source: Federal Emergency Management Agency (FEMA)

Appendix II

Building Vulnerability Assessment Checklist

T he Building Vulnerability Assessment Checklist developed by FEMA can be used as a screening tool for preliminary design vulnerability assessment. In addition to examining design issues that affect vulnerability, the checklist includes questions that determine if critical systems continue to function in order to enhance deterrence, detection, denial and damage limitation, and to ensure that emergency systems function during a threat or hazard situation.

Building Vulnerability Assessment Checklist

Section	Vulnerability Question	Guidance	Observations
1	**Site**		
1.1	What major structures surround the facility (site or building(s))? What critical infrastructure, government, military, or recreation facilities are in the local area that impact transportation, utilities, and collateral damage (attack at this facility impacting the other major structures or attack on the major structures impacting this facility)? What are the adjacent land uses immediately outside the perimeter of this facility (site or building(s))?	**Critical infrastructure to consider includes:** **Telecommunications infrastructure** Facilities for broadcast TV, cable TV; cellular networks; newspaper offices, production, and distribution; radio stations; satellite base stations; telephone trunking and switching stations, including critical cable routes and major rights-of-way **Electric power systems** Power plants, especially nuclear facilities; transmission and distribution system components; fuel distribution, delivery, and storage **Gas and oil facilities** Hazardous material facilities, oil/gas pipelines, and storage facilities	

Building Vulnerability Assessment Checklist (Continued)

Section	Vulnerability Question	Guidance	Observations
	Do future development plans change these land uses outside the facility (site or building (s)) perimeter? Although this question bridges threat and vulnerability, the threat is the manmade hazard that can occur (likelihood and impact) and the vulnerability is the proximity of the hazard to the building(s) being assessed. Thus, a chemical plant release may be a threat/hazard, but vulnerability changes if the plant is 1 mile upwind for the prevailing winds versus 10 miles away and downwind. Similarly, a terrorist attack upon an adjacent building may impact the building(s) being assessed. The Murrah Federal Building in Oklahoma City was not the only building to have severe damage caused by the explosion of the Ryder rental truck bomb.	**Banking and finance institutions** Financial institutions (banks, credit unions) and the business district; note schedule business/financial district may follow; armored car services **Transportation networks** Airports: carriers, flight paths, and airport layout; location of air traffic control towers, runways, passenger terminals, and parking areas Bus Stations Pipelines: oil; gas Trains/Subways: rails and lines, railheads/rail yards, interchanges, tunnels, and cargo/passenger terminals; note hazardous material transported Traffic: interstate highways/roads/tunnels/bridges carrying large volumes; points of congestion; note time of day and day of week Trucking: hazardous materials cargo loading/unloading facilities; truck terminals, weigh stations, and rest areas Waterways: dams; levees; berths and ports for cruise ships, ferries, roll-on/roll-off cargo vessels, and container ships; international (foreign) flagged vessels (and cargo) **Water supply systems** Pipelines and process/treatment facilities, dams for water collection; wastewater treatment **Government services** Federal/state/local government offices – post offices, law enforcement stations, fire/rescue, town/city hall, local mayor's/governor's residences, judicial offices and courts, military installations (include type-Active, Reserves, National Guard) **Emergency services** Backup facilities, communications centers, Emergency Operations Centers (EOCs), fire/Emergency Medical Service (EMS) facilities, Emergency Medical Center (EMCs), law enforcement facilities	

ASSET VALUE, THREAT/HAZARD, VULNERABILITY, AND RISK

Building Vulnerability Assessment Checklist (Continued)

Section	Vulnerability Question	Guidance	Observations
		The following are not critical infrastructure, but have potential collateral damage to consider: **Agricultural facilities:** chemical distribution, storage, and application sites; crop spraying services; farms and ranches; food processing, storage, and distribution facilities **Commercial/manufacturing/industrial facilities:** apartment buildings; business/corporate centers; chemical plants (especially those with Section 302 Extremely Hazardous Substances); factories; fuel production, distribution, and storage facilities; hotels and convention centers; industrial plants; raw material production, distribution, and storage facilities; research facilities and laboratories; shipping, warehousing, transfer, and logistical centers **Events and attractions:** festivals and celebrations; open-air markets; parades; rallies, demonstrations, and marches; religious services; scenic tours; theme parks **Health care system components:** family planning clinics; health department offices; hospitals; radiological material and medical waste transportation, storage, and disposal; research facilities and laboratories, walk-in clinics **Political or symbolically significant sites:** embassies, consulates, landmarks, monuments, political party and special interest groups offices, religious sites **Public/private institutions:** academic institutions, cultural centers, libraries, museums, research facilities and laboratories, schools **Recreation facilities:** auditoriums, casinos, concert halls and pavilions, parks, restaurants and clubs (frequented by potential target populations), sports arenas, stadiums, theaters, malls, and special interest group facilities; note congestion dates and times for shopping centers References: *FEMA 386-7, FEMA SLG 101, DOJ NCJ181200*	
1.2	Does the terrain place the building in a depression or low area?	Depressions or low areas can trap heavy vapors, inhibit natural decontamination by prevailing winds, and reduce the effectiveness of in-place sheltering. Reference: *USAF Installation Force Protection Guide*	

ASSET VALUE, THREAT/HAZARD, VULNERABILITY, AND RISK

Building Vulnerability Assessment Checklist (Continued)

Section	Vulnerability Question	Guidance	Observations
1.3	In dense, urban areas, does curb lane parking allow uncontrolled vehicles to park unacceptably close to a building in public rights-of-way?	Where distance from the building to the nearest curb provides insufficient setback, restrict parking in the curb lane. For typical city streets, this may require negotiating to close the curb lane. Setback is common terminology for the distance between a building and its associated roadway or parking. It is analogous to stand-off between a vehicle bomb and the building. The benefit per foot of increased stand-off between a potential vehicle bomb and a building is very high when close to a building and decreases rapidly as the distance increases. Note that the July 1, 1994, Americans with Disabilities Act Standards for Accessible Design states that required handicapped parking shall be located on the shortest accessible route of travel from adjacent parking to an accessible entrance. Reference: *GSA PBS-P100*	
1.4	Is a perimeter fence or other types of barrier controls in place?	The intent is to channel pedestrian traffic onto a site with multiple buildings through known access control points. For a single building, the intent is to have a single visitor entrance. Reference: *GSA PBS-P100*	
1.5	What are the site access points to the site or building?	The goal is to have at least two access points — one for passenger vehicles and one for delivery trucks due to the different procedures needed for each. Having two access points also helps if one of the access points becomes unusable, then traffic can be routed through the other access point. Reference: *USAF Installation Force Protection Guide*	
1.6	Is vehicle traffic separated from pedestrian traffic on the site?	Pedestrian access should not be endangered by car traffic. Pedestrian access, especially from public transportation, should not cross vehicle traffic if possible. References: *GSA PBS-P100 and FEMA 386-7*	
1.7	Is there vehicle and pedestrian access control at the perimeter of the site?	Vehicle and pedestrian access control and inspection should occur as far from facilities as possible (preferably at the site perimeter) with the ability to regulate the flow of people and vehicles one at a time. Control on-site parking with identification checks, security personnel, and access control systems. Reference: *FEMA 386-7*	

Building Vulnerability Assessment Checklist (Continued)

Section	Vulnerability Question	Guidance	Observations
1.8	Is there space for inspection at the curb line or outside the protected perimeter? What is the minimum distance from the inspection location to the building?	Design features for the vehicular inspection point include: vehicle arrest devices that prevent vehicles from leaving the vehicular inspection area and prevent tailgating. If screening space cannot be provided, consider other design features such as: hardening and alternative location for vehicle search/inspection. Reference: *GSA PBS-P100*	
1.9	Is there any potential access to the site or building through utility paths or water runoff?	Eliminate potential site access through utility tunnels, corridors, manholes, stormwater runoff culverts, etc. Ensure covers to these access points are secured. Reference: *USAF Installation Force Protection Guide*	
1.10	What are the existing types of vehicle anti-ram devices for the site or building? Are these devices at the property boundary or at the building?	Passive barriers include bollards, walls, hardened fences (steel cable interlaced), trenches, ponds/basins, concrete planters, street furniture, plantings, trees, sculptures, and fountains. Active barriers include pop-up bollards, swing arm gates, and rotating plates and drums, etc. Reference: *GSA PBS-P100*	
1.11	What is the anti-ram buffer zone stand-off distance from the building to unscreened vehicles or parking?	If the recommended distance for the postulated threat is not available, consider reducing the stand-off required through structural hardening or manufacturing additional stand-off through barriers and parking restrictions. Also, consider relocation of vulnerable functions within the building, or to a more hazard-resistant building. More stand-off should be used for unscreened vehicles than for screened vehicles that have been searched. Reference: *GSA PBS-P100*	
1.12	Are perimeter barriers capable of stopping vehicles? Will the vehicle barriers at the perimeter and building maintain access for emergency responders, including large fire apparatus?	Anti-ram protection may be provided by adequately designed: bollards, street furniture, sculpture, landscaping, walls, and fences. The anti-ram protection must be able to stop the threat vehicle size (weight) at the speed attainable by that vehicle at impact. If the anti-ram protection cannot absorb the desired kinetic energy, consider adding speed controls (serpentines or speed bumps) to limit the speed at impact. If the resultant speed is still too great, the anti-ram protection should be improved. References: *Military Handbook 1013/14 and GSA PBS P-100*	

Building Vulnerability Assessment Checklist (Continued)

Section	Vulnerability Question	Guidance	Observations
1.13	Does site circulation prevent high-speed approaches by vehicles?	The intent is to use site circulation to minimize vehicle speeds and eliminate direct approaches to structures. Reference: *GSA PBS-P100*	
1.14	Are there offsetting vehicle entrances from the direction of a vehicle's approach to force a reduction of speed?	Single or double 90-degree turns effectively reduce vehicle approach speed. Reference: *GSA PBS-P100*	
1.15	Is there a minimum setback distance between the building and parked vehicles?	Adjacent public parking should be directed to more distant or better-protected areas, segregated from employee parking and away from the building. Some publications use the term setback in lieu of the term stand-off. Reference: *GSA PBS-P100*	
1.16	Does adjacent surface parking on site maintain a minimum stand-off distance?	The specific stand-off distance needed is based upon the design basis threat bomb size and the building construction. For initial screening, consider using 25 meters (82 feet) as a minimum, with more distance needed for unreinforced masonry or wooden walls. Reference: *GSA PBS-P100*	
1.17	Do standalone, aboveground parking garages provide adequate visibility across as well as into and out of the parking garage?	Pedestrian paths should be planned to concentrate activity to the extent possible. Limiting vehicular entry/exits to a minimum number of locations is beneficial. Stair tower and elevator lobby design should be as open as code permits. Stair and/or elevator waiting areas should be as open to the exterior and/or the parking areas as possible and well lighted. Impact-resistant, laminated glass for stair towers and elevators is a way to provide visual openness. Potential hiding places below stairs should be closed off; nooks and crannies should be avoided, and dead-end parking areas should be eliminated. Reference: *GSA PBS-P100*	
1.18	Are garage or service area entrances for employee-permitted vehicles protected by suitable anti-ram devices? Coordinate this protection with other anti-ram devices, such as on the perimeter or property boundary to avoid duplication of arresting capability.	Control internal building parking, underground parking garages, and access to service areas and loading docks in this manner with proper access control, or eliminate the parking altogether. The anti-ram device must be capable of arresting a vehicle of the designated threat size at the speed attainable at the location. Reference: *GSA PBS-P100*	

Building Vulnerability Assessment Checklist (Continued)

Section	Vulnerability Question	Guidance	Observations
1.19	Do site landscaping and street furniture provide hiding places?	Minimize concealment opportunities by keeping landscape plantings (hedges, shrubbery, and large plants with heavy ground cover) and street furniture (bus shelters, benches, trash receptacles, mailboxes, newspaper vending machines) away from the building to permit observation of intruders and prevent hiding of packages. If mail or express boxes are used, the size of the openings should be restricted to prohibit the insertion of packages. Reference: *GSA PBS-P100*	
1.20	Is the site lighting adequate from a security perspective in roadway access and parking areas?	Security protection can be successfully addressed through adequate lighting. The type and design of lighting, including illumination levels, is critical. Illuminating Engineering Society of North America (IESNA) guidelines can be used. The site lighting should be coordinated with the CCTV system. Reference: *GSA PBS-P100*	
1.21	Are line-of-sight perspectives from outside the secured boundary to the building and on the property along pedestrian and vehicle routes integrated with landscaping and green space?	The goal is to prevent the observation of critical assets by persons outside the secure boundary of the site. For individual buildings in an urban environment, this could mean appropriate window treatments or no windows for portions of the building. Once on the site, the concern is to ensure observation by a general workforce aware of any pedestrians or vehicles outside normal circulation routes or attempting to approach the building unobserved. Reference: *USAF Installation Force Protection Guide*	
1.22	Do signs provide control of vehicles and people?	The signage should be simple and have the necessary level of clarity. However, signs that identify sensitive areas should generally not be provided. Reference: *GSA PBS-P100*	
1.23	Are all existing fire hydrants on the site accessible?	Just as vehicle access points to the site must be able to transit emergency vehicles, so too must the emergency vehicles have access to the buildings and, in the case of fire trucks, the fire hydrants. Thus, security considerations must accommodate emergency response requirements. Reference: *GSA PBS-P100*	

Building Vulnerability Assessment Checklist (Continued)

Section	Vulnerability Question	Guidance	Observations
2	**Architectural**		
2.1	Does the site and architectural design incorporate strategies from a Crime Prevention Through Environmental Design (CPTED) perspective?	The focus of CPTED is on creating defensible space by employing: **1. Natural access controls:** – Design streets, sidewalks, and building entrances to clearly indicate public routes and direct people away from private/restricted areas – Discourage access to private areas with structural elements and limit access (no cut-through streets) – Loading zones should be separate from public parking **2. Natural surveillance:** – Design that maximizes visibility of people, parking areas, and building entrances; doors and windows that look out on to streets and parking areas – Shrubbery under 2 feet in height for visibility – Lower branches of existing trees kept at least 10 feet off the ground – Pedestrian-friendly sidewalks and streets to control pedestrian and vehicle circulation – Adequate nighttime lighting, especially at exterior doorways **3. Territorial reinforcement:** – Design that defines property lines – Design that distinguishes private/restricted spaces from public spaces using separation, landscape plantings; pavement designs (pathway and roadway placement); gateway treatments at lobbies, corridors, and door placement; walls, barriers, signage, lighting, and "CPTED" fences – "Traffic-calming" devices for vehicle speed control **4. Target hardening:** – Prohibit entry or access: window locks, deadbolts for doors, interior door hinges – Access control (building and employee/visitor parking) and intrusion detection systems **5. Closed circuit television cameras:** – Prevent crime and influence positive behavior, while enhancing the intended uses of space. In other words, design that eliminates or reduces criminal behavior and at the same time encourages people to "keep an eye out" for each other. References: *GSA PBS-P100 and FEMA 386-7*	

ASSET VALUE, THREAT/HAZARD, VULNERABILITY, AND RISK

Building Vulnerability Assessment Checklist (Continued)

Section	Vulnerability Question	Guidance	Observations
2.2	Is it a mixed-tenant building?	Separate high-risk tenants from low-risk tenants and from publicly accessible areas. Mixed uses may be accommodated through such means as separating entryways, controlling access, and hardening shared partitions, as well as through special security operational countermeasures. Reference: *GSA PBS-P100*	
2.3	Are pedestrian paths planned to concentrate activity to aid in detection?	Site planning and landscape design can provide natural surveillance by concentrating pedestrian activity, limiting entrances/exits, and eliminating concealment opportunities. Also, prevent pedestrian access to parking areas other than via established entrances. Reference: *GSA PBS-P100*	
2.4	Are there trash receptacles and mailboxes in close proximity to the building that can be used to hide explosive devices?	The size of the trash receptacles and mailbox openings should be restricted to prohibit insertion of packages. Street furniture, such as newspaper vending machines, should be kept sufficient distance (10 meters or 33 feet) from the building, or brought inside to a secure area. References: *USAF Installation Force Protection Guide and DoD UCF 4-010-01*	
2.5	Do entrances avoid significant queuing?	If queuing will occur within the building footprint, the area should be enclosed in blast-resistant construction. If queuing is expected outside the building, a rain cover should be provided. For manpower and equipment requirements, collocate or combine staff and visitor entrances. Reference: *GSA PBS-P100*	
2.6	Does security screening cover all public and private areas? Are public and private activities separated? Are public toilets, service spaces, or access to stairs or elevators located in any non-secure areas, including the queuing area before screening at the public entrance?	Retail activities should be prohibited in non-secured areas. However, the Public Building Cooperative Use Act of 1976 encourages retail and mixed uses to create open and inviting buildings. Consider separating entryways, controlling access, hardening shared partitions, and special security operational countermeasures. References: *GSA PBS-P100 and FEMA 386-7*	

Building Vulnerability Assessment Checklist (Continued)

Section	Vulnerability Question	Guidance	Observations
2.7	Is access control provided through main entrance points for employees and visitors? (lobby receptionist, sign-in, staff escorts, issue of visitor badges, checking forms of personal identification, electronic access control systems)	Reference: *Physical Security Assessment for the Department of Veterans Affairs Facilities*	
2.8	Is access to private and public space or restricted area space clearly defined through the design of the space, signage, use of electronic security devices, etc.?	Finishes and signage should be designed for visual simplicity. Reference: *Physical Security Assessment for the Department of Veterans Affairs Facilities*	
2.9	Is access to elevators distinguished as to those that are designated only for employees and visitors?	Reference: *Physical Security Assessment for the Department of Veterans Affairs Facilities*	
2.10	Do public and employee entrances include space for possible future installation of access control and screening equipment?	These include walk-through metal detectors and x-ray devices, identification check, electronic access card, search stations, and turnstiles. Reference: *GSA PBS-P100*	
2.11	Do foyers have reinforced concrete walls and offset interior and exterior doors from each other?	Consider for exterior entrances to the building or to access critical areas within the building if explosive blast hazard must be mitigated. Reference: *U.S. Army TM 5-853*	
2.12	Do doors and walls along the line of security screening meet requirements of UL752 "Standard for Safety: Bullet-Resisting Equipment"?	If the postulated threat in designing entrance access control includes rifles, pistols, or shotguns, then the screening area should have bullet-resistance to protect security personnel and uninvolved bystanders. Glass, if present, should also be bullet-resistant. Reference: *GSA PBS-P100*	
2.13	Do circulation routes have unobstructed views of people approaching controlled access points?	This applies to building entrances and to critical areas within the building. References: *USAF Installation Force Protection Guide and DoD UFC 4-010-01*	

Building Vulnerability Assessment Checklist (Continued)

Section	Vulnerability Question	Guidance	Observations
2.14	Is roof access limited to authorized personnel by means of locking mechanisms?	References: *GSA PBS-P100 and CDC/NIOSH, Pub 2002-139*	
2.15	Are critical assets (people, activities, building systems and components) located close to any main entrance, vehicle circulation, parking, maintenance area, loading dock, or interior parking? Are the critical building systems and components hardened?	Critical building components include: Emergency generator including fuel systems, day tank, fire sprinkler, and water supply; Normal fuel storage; Main switchgear; Telephone distribution and main switchgear; Fire pumps; Building control centers; Uninterruptible Power Supply (UPS) systems controlling critical functions; Main refrigeration and ventilation systems if critical to building operation; Elevator machinery and controls; Shafts for stairs, elevators, and utilities; Critical distribution feeders for emergency power. Evacuation and rescue require emergency systems to remain operational during a disaster and they should be located away from potential attack locations. Primary and backup systems should be separated to reduce the risk of both being impacted by a single incident if collocated. Utility systems should be located at least 50 feet from loading docks, front entrances, and parking areas. One way to harden critical building systems and components is to enclose them within hardened walls, floors, and ceilings. Do not place them near high-risk areas where they can receive collateral damage. Reference: *GSA PBS-P100*	
2.16	Are high-value or critical assets located as far into the interior of the building as possible and separated from the public areas of the building?	Critical assets, such as people and activities, are more vulnerable to hazards when on an exterior building wall or adjacent to uncontrolled public areas inside the building. Reference: *GSA PBS-P100*	
2.17	Is high visitor activity away from critical assets?	High-risk activities should also be separated from low-risk activities. Also, visitor activities should be separated from daily activities. Reference: *USAF Installation Force Protection Guide*	
2.18	Are critical assets located in spaces that are occupied 24 hours per day? Are assets located in areas where they are visible to more than one person?	Reference: *USAF Installation Force Protection Guide*	

Building Vulnerability Assessment Checklist (Continued)

Section	Vulnerability Question	Guidance	Observations
2.19	Are loading docks and receiving and shipping areas separated in any direction from utility rooms, utility mains, and service entrances, including electrical, telephone/data, fire detection/alarm systems, fire suppression water mains, cooling and heating mains, etc.?	Loading docks should be designed to keep vehicles from driving into or parking under the building. If loading docks are in close proximity to critical equipment, consider hardening the equipment and service against explosive blast. Consider a 50-foot separation distance in all directions. Reference: *GSA PBS-P100*	
2.20	Are mailrooms located away from building main entrances, areas containing critical services, utilities, distribution systems, and important assets? Is the mailroom located near the loading dock?	The mailroom should be located at the perimeter of the building with an outside wall or window designed for pressure relief. By separating the mailroom and the loading dock, the collateral damage of an incident at one has less impact upon the other. However, this may be the preferred mailroom location. Off-site screening stations or a separate delivery processing building on site may be cost-effective, particularly if several buildings may share one mailroom. A separate delivery processing building reduces risk and simplifies protection measures. Reference: *GSA PBS-P100*	
2.21	Does the mailroom have adequate space available for equipment to examine incoming packages and for an explosive disposal container?	Screening of all deliveries to the building, including U.S. mail, commercial package delivery services, delivery of office supplies, etc. Reference: *GSA PBS-P100*	
2.22	Are areas of refuge identified, with special consideration given to egress?	Areas of refuge can be safe havens, shelters, or protected spaces for use during specified hazards. Reference: *FEMA 386-7*	
2.23	Are stairwells required for emergency egress located as remotely as possible from high-risk areas where blast events might occur? Are stairways maintained with positive pressure or are there other smoke control systems?	Consider designing stairs so that they discharge into areas other than lobbies, parking, or loading docks. Maintaining positive pressure from a clean source of air (may require special filtering) aids in egress by keeping smoke, heat, toxic fumes, etc,. out of the stairway. Pressurize exit stairways in accordance with the National Model Building Code. References: *GSA PBS-P100* and *CDC/NIOSH, Pub 2002-139*	

ASSET VALUE, THREAT/HAZARD, VULNERABILITY, AND RISK

Building Vulnerability Assessment Checklist (Continued)

Section	Vulnerability Question	Guidance	Observations
2.24	Are enclosures for emergency egress hardened to limit the extent of debris that might otherwise impede safe passage and reduce the flow of evacuees?	Egress pathways should be hardened and discharge into safe areas. Reference: *FEMA 386-7*	
2.25	Do interior barriers differentiate level of security within a building?	Reference: *USAF Installation Force Protection Guide*	
2.26	Are emergency systems located away from high-risk areas?	The intent is to keep the emergency systems out of harm's way, such that one incident does not take out all capability — both the regular systems and their backups. Reference: *FEMA 386-7*	
2.27	Is interior glazing near high-risk areas minimized? Is interior glazing in other areas shatter-resistant?	Interior glazing should be minimized where a threat exists and should be avoided in enclosures of critical functions next to high-risk areas. Reference: *GSA PBS-P100*	
2.28	Are ceiling and lighting systems designed to remain in place during hazard events?	When an explosive blast shatters a window, the blast wave enters the interior space, putting structural and non-structural building components under loads not considered in standard building codes. It has been shown that connection criteria for these systems in high seismic activity areas resulted in much less falling debris that could injure building occupants. Mount all overhead utilities and other fixtures weighing 14 kilograms (31 pounds) or more to minimize the likelihood that they will fall and injure building occupants. Design all equipment mountings to resist forces of 0.5 times the equipment weight in any direction and 1.5 times the equipment weight in the downward direction. This standard does not preclude the need to design equipment mountings for forces required by other criteria, such as seismic standards. Reference: *DoD UCF 4-101-01*	

Building Vulnerability Assessment Checklist (Continued)

Section	Vulnerability Question	Guidance	Observations
3	**Structural Systems**		
3.1	What type of construction? What type of concrete and reinforcing steel? What type of steel? What type of foundation?	The type of construction provides an indication of the robustness to abnormal loading and load reversals. A reinforced concrete moment-resisting frame provides greater ductility and redundancy than a flat-slab or flat-plate construction. The ductility of steel frame with metal deck depends on the connection details and pre-tensioned or post-tensioned construction provides little capacity for abnormal loading patterns and load reversals. The resistance of load-bearing wall structures varies to a great extent, depending on whether the walls are reinforced or un-reinforced. A rapid screening process developed by FEMA for assessing structural hazards identifies the following types of construction with a structural score ranging from 1.0 to 8.5. A higher score indicates a greater capacity to sustain load reversals. Wood buildings of all types - 4.5 to 8.5 Steel moment-resisting frames - 3.5 to 4.5 Braced steel frames - 2.5 to 3.0 Light metal buildings - 5.5 to 6.5 Steel frames with cast-in-place concrete shear walls - 3.5 to 4.5 Steel frames with unreinforced masonry infill walls - 1.5 to 3.0 Concrete moment-resisting frames - 2.0 to 4.0 Concrete shear wall buildings - 3.0 to 4.0 Concrete frames with unreinforced masonry infill walls - 1.5 to 3.0 Tilt-up buildings - 2.0 to 3.5 Precast concrete frame buildings - 1.5 to 2.5 Reinforced masonry - 3.0 to 4.0 Unreinforced masonry - 1.0 to 2.5 References: *FEMA 154 and Physical Security Assessment for the Department of Veterans Affairs Facilities*	
3.2	Do the reinforced concrete structures contain symmetric steel reinforcement (positive and negative faces) in all floor slabs, roof slabs, walls, beams, and girders that may be subjected to rebound, uplift, and suction pressures?	Reference: *GSA PBS-P100*	

ASSET VALUE, THREAT/HAZARD, VULNERABILITY, AND RISK

Building Vulnerability Assessment Checklist (Continued)

Section	Vulnerability Question	Guidance	Observations
	Do the lap splices fully develop the capacity of the reinforcement? Are lap splices and other discontinuities staggered? Do the connections possess ductile details? Is special shear reinforcement, including ties and stirrups, available to allow large post-elastic behavior?		
3.3	Are the steel frame connections moment connections? Is the column spacing minimized so that reasonably sized members will resist the design loads and increase the redundancy of the system? What are the floor-to-floor heights?	A practical upper level for column spacing is generally 30 feet. Unless there is an overriding architectural requirement, a practical limit for floor-to-floor heights is generally less than or equal to 16 feet. Reference: *GSA PBS-P100*	
3.4	Are critical elements vulnerable to failure?	The priority for upgrades should be based on the relative importance of structural or non-structural elements that are essential to mitigating the extent of collapse and minimizing injury and damage. Primary Structural Elements provide the essential parts of the building's resistance to catastrophic blast loads and progressive collapse. These include columns, girders, roof beams, and the main lateral resistance system. Secondary Structural Elements consist of all other load-bearing members, such as floor beams, slabs, etc. Primary Non-Structural Elements consist of elements (including their attachments) that are essential for life safety systems or elements that can cause substantial injury if failure occurs, including ceilings or heavy suspended mechanical units. Secondary Non-Structural Elements consist of all elements not covered in primary non-structural elements, such as partitions, furniture, and light fixtures. Reference: *GSA PBS-P100*	

Building Vulnerability Assessment Checklist (Continued)

Section	Vulnerability Question	Guidance	Observations
3.5	Will the structure suffer an unacceptable level of damage resulting from the postulated threat (blast loading or weapon impact)?	The extent of damage to the structure and exterior wall systems from the bomb threat may be related to a protection level. The following is for new buildings: **Level of Protection Below Antiterrorism Standards** – Severe damage. Frame collapse/massive destruction. Little left standing. Doors and windows fail and result in lethal hazards. Majority of personnel suffer fatalities. **Very Low Level Protection** – Heavy damage. Onset of structural collapse. Major deformation of primary and secondary structural members, but progressive collapse is unlikely. Collapse of non-structural elements. Glazing will break and is likely to be propelled into the building, resulting in serious glazing fragment injuries, but fragments will be reduced. Doors may be propelled into rooms, presenting serious hazards. Majority of personnel suffer serious injuries. There are likely to be a limited number (10 percent to 25 percent) of fatalities. **Low Level of Protection** – Moderate damage, unrepairable. Major deformation of non-structural elements and secondary structural members and minor deformation of primary structural members, but progressive collapse is unlikely. Glazing will break, but fall within 1 meter of the wall or otherwise not present a significant fragment hazard. Doors may fail, but they will rebound out of their frames, presenting minimal hazards. Majority of personnel suffer significant injuries. There may be a few (<10 percent) fatalities. **Medium Level Protection** – Minor damage, repairable. Minor deformations of non-structural elements and secondary structural members and no permanent deformation in primary structural members. Glazing will break, but will remain in the window frame. Doors will stay in frames, but will not be reusable. Some minor injuries, but fatalities are unlikely. **High Level Protection** – Minimal damage, repairable. No permanent deformation of primary and secondary structural members or non-structural elements. Glazing will not break. Doors will be reusable. Only superficial injuries are likely. Reference: *DoD UFC 4-010-01*	
3.6	Is the structure vulnerable to progressive collapse? Is the building capable of sustaining the removal of a column for one floor above grade at	Design to mitigate progressive collapse is an independent analysis to determine a system's ability to resist structural collapse upon the loss of a major structural element or the system's ability to resist the loss of a major structural element. Design to mitigate progressive collapse may be based on the methods outlined in ASCE 7-98 (now 7-02). Designers may apply static and/or	

Building Vulnerability Assessment Checklist (Continued)

Section	Vulnerability Question	Guidance	Observations
	the building perimeter without progressive collapse? In the event of an internal explosion in an uncontrolled public ground floor area, does the design prevent progressive collapse due to the loss of one primary column? Do architectural or structural features provide a minimum 6-inch stand-off to the internal columns (primary vertical load carrying members)? Are the columns in the unscreened internal spaces designed for an unbraced length equal to two floors, or three floors where there are two levels of parking?	dynamic methods of analysis to meet this requirement and ultimate load capacities may be assumed in the analyses. Combine structural upgrades for retrofits to existing buildings, such as seismic and progressive collapse, into a single project due to the economic synergies and other cross benefits. Existing facilities may be retrofitted to withstand the design level threat or to accept the loss of a column for one floor above grade at the building perimeter without progressive collapse. Note that collapse of floors or roof must not be permitted. Reference: GSA PBS-P100	
3.7	Are there adequate redundant load paths in the structure?	Special consideration should be given to materials that have inherent ductility and that are better able to respond to load reversals, such as cast in place reinforced concrete, reinforced masonry, and steel construction. Careful detailing is required for material such as pre-stressed concrete, pre-cast concrete, and masonry to adequately respond to the design loads. Primary vertical load carrying members should be protected where parking is inside a facility and the building superstructure is supported by the parking structure. Reference: GSA PBS-P100	
3.8	Are there transfer girders supported by columns within unscreened public spaces or at the exterior of the building?	Transfer girders allow discontinuities in columns between the roof and foundation. This design has inherent difficulty in transferring load to redundant paths upon loss of a column or the girder. Transfer beams and girders that, if lost, may cause progressive collapse are highly discouraged. Reference: GSA PBS-P100	
3.9	What is the grouting and reinforcement of masonry (brick and/or concrete masonry unit (CMU)) exterior walls?	Avoid unreinforced masonry exterior walls. Reinforcement can run the range of light to heavy, depending upon the stand-off distance available and postulated design threat. Reference: GSA PBS-P100 recommends fully grouted and reinforced CMU construction where CMU is selected.	

Building Vulnerability Assessment Checklist (Continued)

Section	Vulnerability Question	Guidance	Observations
		Reference: *DoD Minimum Antiterrorism Standards for Buildings* states "Unreinforced masonry walls are prohibited for the exterior walls of new buildings. A minimum of 0.05 percent vertical reinforcement with a maximum spacing of 1200 mm (48 in) will be provided. For existing buildings, implement mitigating measures to provide an equivalent level of protection." [This is light reinforcement and based upon the recommended stand-off distance for the situation.]	
3.10	Will the loading dock design limit damage to adjacent areas and vent explosive force to the exterior of the building?	Design the floor of the loading dock for blast resistance if the area below is occupied or contains critical utilities. Reference: *GSA PBS-P100*	
3.11	Are mailrooms, where packages are received and opened for inspection, and unscreened retail spaces designed to mitigate the effects of a blast on primary vertical or lateral bracing members?	Where mailrooms and unscreened retail spaces are located in occupied areas or adjacent to critical utilities, walls, ceilings, and floors, they should be blast- and fragment-resistant. Methods to facilitate the venting of explosive forces and gases from the interior spaces to the outside of the structure may include blow-out panels and window system designs that provide protection from blast pressure applied to the outside, but that readily fail and vent if exposed to blast pressure on the inside. Reference: *GSA PBS-P100*	
4	**Building Envelope**		
4.1	What is the designed or estimated protection level of the exterior walls against the postulated explosive threat?	The performance of the façade varies to a great extent on the materials. Different construction includes brick or stone with block backup, steel stud walls, precast panels, or curtain wall with glass, stone, or metal panel elements. Shear walls that are essential to the lateral and vertical load bearing system and that also function as exterior walls should be considered primary structures and should resist the actual blast loads predicted from the threats specified. Where exterior walls are not designed for the full design loads, special consideration should be given to construction types that reduce the potential for injury. Reference: *GSA PBS-P100*	

ASSET VALUE, THREAT/HAZARD, VULNERABILITY, AND RISK

Building Vulnerability Assessment Checklist (Continued)

Section	Vulnerability Question	Guidance	Observations
4.2	Is there less than a 40 percent fenestration opening per structural bay? Is the window system design on the exterior façade balanced to mitigate the hazardous effects of flying glazing following an explosive event? (glazing, frames, anchorage to supporting walls, etc.) Do the glazing systems with a ½-inch (¾-inch is better) bite contain an application of structural silicone? Is the glazing laminated or is it protected with an anti-shatter (fragment retention) film? If an anti-shatter film is used, is it a minimum of a 7-mil thick film, or specially manufactured 4-mil thick film?	The performance of the glass will similarly depend on the materials. Glazing may be single pane or double pane, monolithic or laminated, annealed, heat strengthened or fully tempered. The percent fenestration is a balance between protection level, cost, the architectural look of the building within its surroundings, and building codes. One goal is to keep fenestration to below 40 percent of the building envelope vertical surface area, but the process must balance differing requirements. A blast engineer may prefer no windows; an architect may favor window curtain walls; building codes require so much fenestration per square footage of floor area; fire codes require a prescribed window opening area if the window is a designated escape route; and the building owner has cost concerns. Ideally, an owner would want 100 percent of the glazed area to provide the design protection level against the postulated explosive threat (design basis threat— weapon size at the expected stand-off distance). However, economics and geometry may allow 80 percent to 90 percent due to the statistical differences in the manufacturing process for glass or the angle of incidence of the blast wave upon upper story windows (4th floor and higher). Reference: *GSA PBS-P100*	
4.3	Do the walls, anchorage, and window framing fully develop the capacity of the glazing material selected? Are the walls capable of withstanding the dynamic reactions from the windows? Will the anchorage remain attached to the walls of the building during an explosive event without failure? Is the façade connected to backup block or to the structural frame? Are non-bearing masonry walls reinforced?	Government produced and sponsored computer programs coupled with test data and recognized dynamic structural analysis techniques may be used to determine whether the glazing either survives the specified threats or the post damage performance of the glazing protects the occupants. A breakage probability no higher than 750 breaks per 1,000 may be used when calculating loads to frames and anchorage. The intent is to ensure the building envelope provides relatively equal protection against the postulated explosive threat for the walls and window systems for the safety of the occupants, especially in rooms with exterior walls. Reference: *GSA PBS-P100*	
4.4	Does the building contain ballistic glazing?	Glass-clad polycarbonate or laminated polycarbonate are two types of acceptable glazing material.	

Building Vulnerability Assessment Checklist (Continued)

Section	Vulnerability Question	Guidance	Observations
	Does the ballistic glazing meet the requirements of UL 752 Bullet-Resistant Glazing? Does the building contain security-glazing? Does the security-glazing meet the requirements of ASTM F1233 or UL 972, Burglary Resistant Glazing Material? Do the window assemblies containing forced entry resistant glazing (excluding the glazing) meet the requirements of ASTM F 588?	If windows are upgraded to bullet-resistant, burglar-resistant, or forced entry-resistant, ensure that doors, ceilings, and floors, as applicable, can resist the same for the areas of concern. Reference: *GSA PBS-P100*	
4.5	Do non-window openings, such as mechanical vents and exposed plenums, provide the same level of protection required for the exterior wall?	In-filling of blast over-pressures must be considered through non-window openings such that structural members and all mechanical system mountings and attachments should resist these interior fill pressures. These non-window openings should also be as secure as the rest of the building envelope against forced entry. Reference: *GSA PBS-P100*	
5	**Utility Systems**		
5.1	What is the source of domestic water? (utility, municipal, wells, lake, river, storage tank) Is there a secure alternate drinking water supply?	Domestic water is critical for continued building operation. Although bottled water can satisfy requirements for drinking water and minimal sanitation, domestic water meets many other needs — flushing toilets, building heating and cooling system operation, cooling of emergency generators, humidification, etc. Reference: *FEMA 386-7*	
5.2	Are there multiple entry points for the water supply?	If the building or site has only one source of water entering at one location, the entry point should be secure. Reference: *GSA PBS-P100*	
5.3	Is the incoming water supply in a secure location?	Ensure that only authorized personnel have access to the water supply and its components. Reference: *FEMA 386-7*	
5.4	Does the building or site have storage capacity for domestic water?	Operational facilities will require reliance on adequate domestic water supply. Storage capacity can meet short-term needs and use water trucks to replenish for extended outages.	

ASSET VALUE, THREAT/HAZARD, VULNERABILITY, AND RISK

Building Vulnerability Assessment Checklist (Continued)

Section	Vulnerability Question	Guidance	Observations
	How many gallons of storage capacity are available and how long will it allow operations to continue?	Reference: *Physical Security Assessment for the Department of Veterans Affairs Facilities.*	
5.5	What is the source of water for the fire suppression system? (local utility company lines, storage tanks with utility company backup, lake, or river) Are there alternate water supplies for fire suppression?	The fire suppression system water may be supplied from the domestic water or it may have a separate source, separate storage, or nonpotable alternate sources. For a site with multiple buildings, the concern is that the supply should be adequate to fight the worst case situation according to the fire codes. Recent major construction may change that requirement. Reference: *FEMA 386-7*	
5.6	Is the fire suppression system adequate, code-compliant, and protected (secure location)?	Standpipes, water supply control valves, and other system components should be secure or supervised. Reference: *FEMA 386-7*	
5.7	Do the sprinkler/standpipe interior controls (risers) have fire- and blast-resistant separation? Are the sprinkler and standpipe connections adequate and redundant? Are there fire hydrant and water supply connections near the sprinkler/standpipe connections?	The incoming fire protection water line should be encased, buried, or located 50 feet from high-risk areas. The interior mains should be looped and sectionalized. Reference: *GSA PBS-P100*	
5.8	Are there redundant fire water pumps (e.g., one electric, one diesel)? Are the pumps located apart from each other?	Collocating fire water pumps puts them at risk for a single incident to disable the fire suppression system. References: *GSA PBS-P100 and FEMA 386-7*	
5.9	Are sewer systems accessible? Are they protected or secured?	Sanitary and stormwater sewers should be protected from unauthorized access. The main concerns are backup or flooding into the building, causing a health risk, shorting out electrical equipment, and loss of building use. Reference: *Physical Security Assessment for the Department of Veterans Affairs Facilities*	
5.10	What fuel supplies do the building rely upon for critical operation?	Typically, natural gas, propane, or fuel oil are required for continued operation. Reference: *Physical Security Assessment for the Department of Veterans Affairs Facilities*	

ASSET VALUE, THREAT/HAZARD, VULNERABILITY, AND RISK

Building Vulnerability Assessment Checklist (Continued)

Section	Vulnerability Question	Guidance	Observations
5.11	How much fuel is stored on the site or at the building and how long can this quantity support critical operations? How is it stored? How is it secured?	Fuel storage protection is essential for continued operation. Main fuel storage should be located away from loading docks, entrances, and parking. Access should be restricted and protected (e.g., locks on caps and seals). References: *GSA PBS-P100 and Physical Security Assessment for the Department of Veterans Affairs Facilities*	
5.12	Where is the fuel supply obtained? How is it delivered?	The supply of fuel is dependent on the reliability of the supplier. Reference: *Physical Security Assessment for the Department of Veterans Affairs Facilities*	
5.13	Are there alternate sources of fuel? Can alternate fuels be used?	Critical functions may be served by alternate methods if normal fuel supply is interrupted. Reference: *Physical Security Assessment for the Department of Veterans Affairs Facilities*	
5.14	What is the normal source of electrical service for the site or building?	Utilities are the general source unless co-generation or a private energy provider is available. Reference: *Physical Security Assessment for the Department of Veterans Affairs Facilities*	
5.15	Is there a redundant electrical service source? Can the site or buildings be fed from more than one utility substation?	The utility may have only one source of power from a single substation. There may be only single feeders from the main substation. Reference: *Physical Security Assessment for the Department of Veterans Affairs Facilities*	
5.16	How many service entry points does the site or building have for electricity?	Electrical supply at one location creates a vulnerable situation unless an alternate source is available. Ensure disconnecting requirements according to NFPA 70 (National Fire Protection Association, National Electric Code) are met for multiple service entrances. Reference: *Physical Security Assessment for the Department of Veterans Affairs Facilities*	
5.17	Is the incoming electric service to the building secure?	Typically, the service entrance is a locked room, inaccessible to the public. Reference: *Physical Security Assessment for the Department of Veterans Affairs Facilities*	

Building Vulnerability Assessment Checklist (Continued)

Section	Vulnerability Question	Guidance	Observations
5.18	**What provisions for emergency power exist? What systems receive emergency power and have capacity requirements been tested?** **Is the emergency power collocated with the commercial electric service?** **Is there an exterior connection for emergency power?**	Besides installed generators to supply emergency power, portable generators or rental generators available under emergency contract can be quickly connected to a building with an exterior quick disconnect already installed. Testing under actual loading and operational conditions ensures the critical systems requiring emergency power receive it with a high assurance of reliability. Reference: *GSA PBS-P100*	
5.19	**By what means do the main telephone and data communications interface the site or building?**	Typically, communication ducts or other conduits are available. Overhead service is more identifiable and vulnerable. Reference: *Physical Security Assessment for the Department of Veterans Affairs Facilities*	
5.20	**Are there multiple or redundant locations for the telephone and communications service?**	Secure locations of communications wiring entry to the site or building are required. Reference: *Physical Security Assessment for the Department of Veterans Affairs Facilities*	
5.21	**Does the fire alarm system require communication with external sources?** **By what method is the alarm signal sent to the responding agency: telephone, radio, etc.?** **Is there an intermediary alarm monitoring center?**	Typically, the local fire department responds to an alarm that sounds at the station or is transmitted over phone lines by an auto dialer. An intermediary control center for fire, security, and/or building system alarms may receive the initial notification at an on-site or off-site location. This center may then determine the necessary response and inform the responding agency. Reference: *Physical Security Assessment for the Department of Veterans Affairs Facilities*	
5.22	**Are utility lifelines aboveground, underground, or direct buried?**	Utility lifelines (water, power, communications, etc.) can be protected by concealing, burying, or encasing. References: *GSA PBS-P100 and FEMA 386-7*	

Building Vulnerability Assessment Checklist (Continued)

Section	Vulnerability Question	Guidance	Observations
6	**Mechanical Systems (HVAC and CBR)**		
6.1	Where are the air intakes and exhaust louvers for the building? (low, high, or midpoint of the building structure) Are the intakes and exhausts accessible to the public?	Air intakes should be located on the roof or as high as possible. Otherwise secure within CPTED-compliant fencing or enclosure. The fencing or enclosure should have a sloped roof to prevent the throwing of anything into the enclosure near the intakes. Reference: *GSA PBS-P100* states that air intakes should be on the fourth floor or higher and, on buildings with three floors or less, they should be on the roof or as high as practical. Locating intakes high on a wall is preferred over a roof location. Reference: *DoD UFC 4-010-01* states that, for all new inhabited buildings covered by this document, all air intakes should be located at least 3 meters (10 feet) above the ground. Reference: *CDC/NIOSH, Pub 2002-139* states: "An extension height of 12 feet (3.7 m) will place the intake out of reach of individuals without some assistance. Also, the entrance to the intake should be covered with a sloped metal mesh to reduce the threat of objects being tossed into the intake. A minimum slope of 45° is generally adequate. Extension height should be increased where existing platforms or building features (i.e., loading docks, retaining walls) might provide access to the outdoor air intakes". Reference: *LBNL PUB-51959:* Exhausts are also a concern during an outdoor release, especially if exhaust fans are not in continuous operation, due to wind effects and chimney effects (air movement due to differential temperature).	
6.2	Is roof access limited to authorized personnel by means of locking mechanisms? Is access to mechanical areas similarly controlled?	Roofs are like entrances to the building and are like mechanical rooms when HVAC is installed. Adjacent structures or landscaping should not allow access to the roof. References: *GSA PBS-P100, CDC/NIOSH Pub 2002-139, and LBNL Pub 51959*	
6.3	Are there multiple air intake locations?	Single air intakes may feed several air handling units. Indicate if the air intakes are localized or separated. Installing low-leakage dampers is one way to provide the system separation when necessary. Reference: *Physical Security Assessment for the Department of Veterans Affairs Facilities*	
6.4	What are the types of air filtration? Include the efficiency and number of filter modules for	MERV – Minimum Efficiency Reporting Value HEPA – High Efficiency Particulate Air	

Building Vulnerability Assessment Checklist (Continued)

Section	Vulnerability Question	Guidance	Observations
	each of the main air handling systems? **Is there any collective protection for chemical, biological, and radiological contamination designed into the building?**	Activated charcoal for gases Ultraviolet C for biologicals Consider mix of approaches for optimum protection and cost-effectiveness. Reference: *CDC/NIOSH Pub 2002-139*	
6.5	**Is there space for larger filter assemblies on critical air handling systems?**	Air handling units serving critical functions during continued operation may be retrofitted to provide enhanced protection during emergencies. However, upgraded filtration may have negative effects upon the overall air handling system operation, such as increased pressure drop. Reference: *CDC/NIOSH Pub 2002-139*	
6.6	**Are there μprovisions for air monitors or sensors for chemical or biological agents?**	Duct mounted sensors are usuallly found in limited cases in laboratory areas. Sensors generally have a limited spectrum of high reliability and are costly. Many different technologies are undergoing research to provide capability. Reference: *CDC/NIOSH Pub 2002-139*	
6.7	**By what method are air intakes and exhausts closed when not operational?**	Motorized (low-leakage, fast-acting) dampers are the preferred method for closure with fail-safe to the closed position so as to support in-place sheltering. References: *CDC/NIOSH Pub 2002-139 and LBNL Pub 51959*	
6.8	**How are air handling systems zoned?** **What areas and functions do each of the primary air handling systems serve?**	Understanding the critical areas of the building that must continue functioning focuses security and hazard mitigation measures. Applying HVAC zones that isolate lobbies, mailrooms, loading docks, and other entry and storage areas from the rest of the building HVAC zones and maintaining negative pressure within these areas will contain CBR releases. Identify common return systems that service more than one zone, effectively making a large single zone. Conversely, emergency egress routes should receive positive pressurization to ensure contamination does not hinder egress. Consider filtering of the pressurization air. References: *CDC/NIOSH Pub 2002-139 and LBNL Pub 51959*	
6.9	**Are there large central air handling units or are there multiple units serving separate zones?**	Independent units can continue to operate if damage occurs to limited areas of the building. Reference: *Physical Security Assessment for the Department of Veterans Affairs Facilities*	

Building Vulnerability Assessment Checklist (Continued)

Section	Vulnerability Question	Guidance	Observations
6.10	Are there any redundancies in the air handling system? Can critical areas be served from other units if a major system is disabled?	Redundancy reduces the security measures required compared to a non-redundant situation. Reference: *Physical Security Assessment for the Department of Veterans Affairs Facilities*	
6.11	Is the air supply to critical areas compartmentalized? Similarly, are the critical areas or the building as a whole, considered tight with little or no leakage?	During chemical, biological, and radiological situations, the intent is to either keep the contamination localized in the critical area or prevent its entry into other critical, non-critical, or public areas. Systems can be cross-connected through building openings (doorways, ceilings, partial wall), ductwork leakage, or pressure differences in air handling system. In standard practice, there is almost always some air carried between ventilation zones by pressure imbalances, due to elevator piston action, chimney effect, and wind effects. Smoke testing of the air supply to critical areas may be necessary. Reference: *CDC/NIOSH Pub 2002-139 and LBNL Pub 51959*	
6.12	Are supply, return, and exhaust air systems for critical areas secure? Are all supply and return ducts completely connected to their grilles and registers and secure? Is the return air not ducted?	The air systems to critical areas should be inaccessible to the public, especially if the ductwork runs through the public areas of the building. It is also more secure to have a ducted air handling system versus sharing hallways and plenums above drop ceilings for return air. Non-ducted systems provide greater opportunity for introducing contaminants. Reference: *CDC/NIOSH Pub 2002-139 and LBNL Pub 51959*	
6.13	What is the method of temperature and humidity control? Is it localized or centralized?	Central systems can range from monitoring only to full control. Local control may be available to override central operation. Of greatest concern are systems needed before, during, and after an incident that may be unavailable due to temperature and humidity exceeding operational limits (e.g., main telephone switch room). Reference: *DOC CIAO Vulnerability Assessment Framework 1.1*	
6.14	Where are the building automation control centers and cabinets located?	Access to any component of the building automation and control system could compromise the functioning of the system, increasing vulnerability to a hazard or precluding their proper operation during a hazard incident.	

ASSET VALUE, THREAT/HAZARD, VULNERABILITY, AND RISK

Building Vulnerability Assessment Checklist (Continued)

Section	Vulnerability Question	Guidance	Observations
	Are they in secure areas? How is the control wiring routed?	The HVAC and exhaust system controls should be in a secure area that allows rapid shutdown or other activation based upon location and type of attack. References: *FEMA 386-7, DOC CIAO Vulnerability Assessment Framework 1.1 and LBNL Pub 51959*	
6.15	Does the control of air handling systems support plans for sheltering in place or other protective approach?	The micro-meteorological effects of buildings and terrain can alter travel and duration of chemical agents and hazardous material releases. Shielding in the form of sheltering in place can protect people and property from harmful effects. To support in-place sheltering, the air handling systems require the ability for authorized personnel to rapidly turn off all systems. However, if the system is properly filtered, then keeping the system operating will provide protection as long as the air handling system does not distribute an internal release to other portions of the building. Reference: *CDC/NIOSH Pub 2002-139*	
6.16	Are there any smoke evacuation systems installed? Does it have purge capability?	For an internal blast, a smoke removal system may be essential, particularly in large, open spaces. The equipment should be located away from high-risk areas, the system controls and wiring should be protected, and it should be connected to emergency power. This exhaust capability can be built into areas with significant risk on internal events, such as lobbies, loading docks, and mailrooms. Consider filtering of the exhaust to capture CBR contaminants. References: *GSA PBS-P100, CDC/NIOSH Pub 2002-139, and LBNL Pub 51959*	
6.17	Where is roof-mounted equipment located on the roof? (near perimeter, at center of roof)	Roof-mounted equipment should be kept away from the building perimeter. Reference: *U.S. Army TM 5-853*	
6.18	Are fire dampers installed at all fire barriers? Are all dampers functional and seal well when closed?	All dampers (fire, smoke, outdoor air, return air, bypass) must be functional for proper protection within the building during an incident. Reference: *CDC/NIOSH Pub 2002-139*	
6.19	Do fire walls and fire doors maintain their integrity?	The tightness of the building (both exterior, by weatherization to seal cracks around doors and windows, and internal, by zone ducting, fire walls, fire stops, and fire doors) provides energy conservation benefits and functional benefits during a CBR incident. Reference: *LBNL Pub 51959*	

ASSET VALUE, THREAT/HAZARD, VULNERABILITY, AND RISK

Building Vulnerability Assessment Checklist (Continued)

Section	Vulnerability Question	Guidance	Observations
6.20	Do elevators have recall capability and elevator emergency message capability?	Although a life-safety code and fire response requirement, the control of elevators also has benefit during a CBR incident. The elevators generate a piston effect, causing pressure differentials in the elevator shaft and associated floors that can force contamination to flow up or down. Reference: *LBNL Pub 51959*	
6.21	Is access to building information restricted?	Information on building operations, schematics, procedures, plans, and specifications should be strictly controlled and available only to authorized personnel. References: *CDC/NIOSH Pub 2002-139 and LBNL Pub 51959*	
6.22	Does the HVAC maintenance staff have the proper training, procedures, and preventive maintenance schedule to ensure CBR equipment is functional?	Functional equipment must interface with operational procedures in an emergency plan to ensure the equipment is properly operated to provide the protection desired. The HVAC system can be operated in different ways, depending upon an external or internal release and where in the building an internal release occurs. Thus maintenance and security staff must have the training to properly operate the HVAC system under different circumstances, even if the procedure is to turn off all air movement equipment. Reference: *CDC/NIOSH Pub 2002-139 and LBNL Pub 51959*	
7	**Plumbing and Gas Systems**		
7.1	What is the method of water distribution?	Central shaft locations for piping are more vulnerable than multiple riser locations. Reference: *Physical Security Assessment for the Department of Veterans Affairs Facilities*	
7.2	What is the method of gas distribution? (heating, cooking, medical, process)	Reference: *Physical Security Assessment for the Department of Veterans Affairs Facilities*	
7.3	Is there redundancy to the main piping distribution?	Looping of piping and use of section valves provide redundancies in the event sections of the system are damaged. Reference: *Physical Security Assessment for the Department of Veterans Affairs Facilities*	

ASSET VALUE, THREAT/HAZARD, VULNERABILITY, AND RISK

Building Vulnerability Assessment Checklist (Continued)

Section	Vulnerability Question	Guidance	Observations
7.4	What is the method of heating domestic water? What fuel(s) is used?	Single source of hot water with one fuel source is more vulnerable than multiple sources and multiple fuel types. Domestic hot water availability is an operational concern for many building occupancies. Reference: *Physical Security Assessment for the Department of Veterans Affairs Facilities*	
7.5	Where are gas storage tanks located? (heating, cooking, medical, process) How are they piped to the distribution system? (above or below ground)	The concern is that the tanks and piping could be vulnerable to a moving vehicle or a bomb blast either directly or by collateral damage due to proximity to a higher-risk area. Reference: *Physical Security Assessment for the Department of Veterans Affairs Facilities*	
7.6	Are there reserve supplies of critical gases?	Localized gas cylinders could be available in the event of damage to the central tank system. Reference: *Physical Security Assessment for the Department of Veterans Affairs Facilities*	
8	**Electrical Systems**		
8.1	Are there any transformers or switchgears located outside the building or accessible from the building exterior? Are they vulnerable to public access? Are they secured?	Reference: *Physical Security Assessment for the Department of Veterans Affairs Facilities*	
8.2	What is the extent of the external building lighting in utility and service areas and at normal entryways used by the building occupants?	Reference: *Physical Security Assessment for the Department of Veterans Affairs Facilities*	
8.3	How are the electrical rooms secured and where are they located relative to other higher-risk areas, starting with the main electrical distribution room at the service entrance?	Reference: *Physical Security Assessment for the Department of Veterans Affairs Facilities*	

Building Vulnerability Assessment Checklist (Continued)

Section	Vulnerability Question	Guidance	Observations
8.4	Are critical electrical systems collocated with other building systems? Are critical electrical systems located in areas outside of secured electrical areas? Is security system wiring located separately from electrical and other service systems?	Collocation concerns include rooms, ceilings, raceways, conduits, panels, and risers. Reference: *Physical Security Assessment for the Department of Veterans Affairs Facilities*	
8.5	How are electrical distribution panels serving branch circuits secured or are they in secure locations?	Reference: *Physical Security Assessment for the Department of Veterans Affairs Facilities*	
8.6	Does emergency backup power exist for all areas within the building or for critical areas only? How is the emergency power distributed? Is the emergency power system independent from the normal electrical service, particularly in critical areas?	There should be no single critical node that allows both the normal electrical service and the emergency backup power to be affected by a single incident. Automatic transfer switches and interconnecting switchgear are the initial concerns. Emergency and normal electrical equipment should be installed separately, at different locations, and as far apart as possible. Reference: *GSA PBS-P100*	
8.7	How is the primary electrical system wiring distributed? Is it collocated with other major utilities? Is there redundancy of distribution to critical areas?	Central utility shafts may be subject to damage, especially if there is only one for the building. Reference: *Physical Security Assessment for the Department of Veterans Affairs Facilities*	

ASSET VALUE, THREAT/HAZARD, VULNERABILITY, AND RISK

Building Vulnerability Assessment Checklist (Continued)

Section	Vulnerability Question	Guidance	Observations
9	**Fire Alarm Systems**		
9.1	Is the building fire alarm system centralized or localized? How are alarms made known, both locally and centrally? Are critical documents and control systems located in a secure yet accessible location?	Fire alarm systems must first warn building occupants to evacuate for life safety. Then they must inform the responding agency to dispatch fire equipment and personnel. Reference: *Physical Security Assessment for the Department of Veterans Affairs Facilities*	
9.2	Where are the fire alarm panels located? Do they allow access to unauthorized personnel?	Reference: *Physical Security Assessment for the Department of Veterans Affairs Facilities*	
9.3	Is the fire alarm system standalone or integrated with other functions such as security and environmental or building management systems? What is the interface?	Reference: *Physical Security Assessment for the Department of Veterans Affairs Facilities*	
9.4	Do key fire alarm system components have fire- and blast-resistant separation?	This is especially necessary for the fire command center or fire alarm control center. The concern is to similarly protect critical components as described in Items 2.19, 5.7, and 10.3.	
9.5	Is there redundant off-premises fire alarm reporting?	Fire alarms can ring at a fire station, at an intermediary alarm monitoring center, or autodial someone else. See Items 5.21 and 10.5.	
10	**Communications and IT Systems**		
10.1	Where is the main telephone distribution room and where is it in relation to higher-risk areas? Is the main telephone distribution room secure?	One can expect to find voice, data, signal, and alarm systems to be routed through the main telephone distribution room. Reference: *FEMA 386-7*	

Building Vulnerability Assessment Checklist (Continued)

Section	Vulnerability Question	Guidance	Observations
10.2	Does the telephone system have an uninterruptible power supply (UPS)? What is its type, power rating, and operational duration under load, and location? (battery, on-line, filtered)	Many telephone systems are now computerized and need a UPS to ensure reliability during power fluctuations. The UPS is also needed to await any emergency power coming on line or allow orderly shutdown. Reference: *DOC CIAO Vulnerability Assessment Framework 1.1*	
10.3	Where are communication systems wiring closets located? (voice, data, signal, alarm) Are they collocated with other utilities? Are they in secure areas?	Concern is to have separation distance from other utilities and higher-risk areas to avoid collateral damage. Security approaches on the closets include door alarms, closed circuit television, swipe cards, or other logging notifications to ensure only authorized personnel have access to these closets. Reference: *FEMA 386-7*	
10.4	How is the communications system wiring distributed? (secure chases and risers, accessible public areas)	The intent is to prevent tampering with the systems. Reference: *Physical Security Assessment for the Department of Veterans Affairs Facilities*	
10.5	Are there redundant communications systems available?	Critical areas should be supplied with multiple or redundant means of communications. Power outage phones can provide redundancy as they connect directly to the local commercial telephone switch off site and not through the building telephone switch in the main telephone distribution room. A base radio communication system with antenna can be installed in stairwells, and portable sets distributed to floors. References: *GSA PBS-P100 and FEMA 386-7*	
10.6	Where are the main distribution facility, data centers, routers, firewalls, and servers located and are they secure? Where are the secondary and/or intermediate distribution facilities and are they secure?	Concern is collateral damage from manmade hazards and redundancy of critical functions. Reference: *DOC CIAO Vulnerability Assessment Framework 1.1*	
10.7	What type and where are the Wide Area Network (WAN) connections?	Critical facilities should have two Minimum-Points-of-Presence(MPOPs) where the telephone company's outside cable terminates inside the building. It is functionally a service entrance connection that demarcates where the telephone company's property stops and the building owner's property begins. The MPOPs should not be	

ASSET VALUE, THREAT/HAZARD, VULNERABILITY, AND RISK

Building Vulnerability Assessment Checklist (Continued)

Section	Vulnerability Question	Guidance	Observations
		collocated and they should connect to different telephone company central offices so that the loss of one cable or central office does not reduce capability. Reference: *Physical Security Assessment for the Department of Veterans Affairs Facilities*	
10.8	**What are the type, power rating, and location of the uninterruptible power supply?** (battery, on-line, filtered) **Are the UPS also connected to emergency power?**	Consider that UPS should be found at all computerized points from the main distribution facility to individual data closets and at critical personal computers/terminals. Critical LAN sections should also be on backup power. Reference: *DOC CIAO Vulnerability Assessment Framework 1.1*	
10.9	**What type of Local Area Network (LAN) cabling and physical topology is used?** (Category (Cat) 5, Gigabit Ethernet, Ethernet, Token Ring)	The physical topology of a network is the way in which the cables and computers are connected to each other. The main types of physical topologies are: Bus (single radial where any damage on the bus affects the whole system, but especially all portions downstream) Star (several computes are connected to a hub and many hubs can be in the network — the hubs can be critical nodes, but the other hubs continue to function if one fails) Ring (a bus with a continuous connection - least used, but can tolerate some damage because if the ring fails at a single point it can be rerouted much like a looped electric or water system) The configuration and the availability of surplus cable or spare capacity on individual cables can reduce vulnerability to hazard incidents. Reference: *Physical Security Assessment for the Department of Veterans Affairs Facilities*	
10.10	**For installed radio/wireless systems, what are their types and where are they located?** (radio frequency (RF), high frequency (HF), very high frequency (VHF), medium wave (MW))	Depending upon the function of the wireless system, it could be susceptible to accidental or intended jamming or collateral damage. Reference: *Physical Security Assessment for the Department of Veterans Affairs Facilities*	
10.11	**Do the Information Technology (IT - computer) systems meet requirements of confidentiality, integrity, and availability?**	Ensure access to terminals and equipment for authorized personnel only and ensure system up-time to meet operational needs. Reference: *DOC CIAO Vulnerability Assessment Framework 1.1*	

ASSET VALUE, THREAT/HAZARD, VULNERABILITY, AND RISK

Building Vulnerability Assessment Checklist (Continued)

Section	Vulnerability Question	Guidance	Observations
10.12	Where is the disaster recovery/ mirroring site?	A site with suitable equipment that allows continuation of operations or that mirrors (operates in parallel to) the existing operation is beneficial if equipment is lost during a natural or manmade disaster. The need is based upon the criticality of the operation and how quickly replacement equipment can be put in place and operated. Reference: *DOC CIAO Vulnerability Assessment Framework 1.1*	
10.13	Where is the backup tape/file storage site and what is the type of safe environment? (safe, vault, underground) Is there redundant refrigeration in the site?	If equipment is lost, data are most likely lost, too. Backups are needed to continue operations at the disaster recovery site or when equipment can be delivered and installed. Reference: *DOC CIAO Vulnerability Assessment Framework 1.1*	
10.14	Are there any satellite communications (SATCOM) links? (location, power, UPS, emergency power, spare capacity/capability)	SATCOM links can serve as redundant communications for voice and data if configured to support required capability after a hazard incident. Reference: *DOC CIAO Vulnerability Assessment Framework 1.1*	
10.15	Is there a mass notification system that reaches all building occupants? (public address, pager, cell phone, computer override, etc.) Will one or more of these systems be operational under hazard conditions? (UPS, emergency power)	Depending upon building size, a mass notification system will provide warning and alert information, along with actions to take before and after an incident if there is redundancy and power. Reference: *DoD UFC 4-010-01*	
10.16	Do control centers and their designated alternate locations have equivalent or reduced capability for voice, data, mass notification, etc.? (emergency operations, security, fire alarms, building automation) Do the alternate locations also have access to backup systems, including emergency power?	Reference: *GSA PBS-P100*	

ASSET VALUE, THREAT/HAZARD, VULNERABILITY, AND RISK

Building Vulnerability Assessment Checklist (Continued)

Section	Vulnerability Question	Guidance	Observations
11	**Equipment Operations and Maintenance**		
11.1	Are there composite drawings indicating location and capacities of major systems and are they current? (electrical, mechanical, and fire protection; and date of last update)	Within critical infrastructure protection at the building level, the current configuration and capacity of all critical systems must be understood to ensure they meet emergency needs. Manuals must also be current to ensure operations and maintenance keeps these systems properly functioning. The system must function during an emergency unless directly affected by the hazard incident.	
	Do updated operations and maintenance (O&M) manuals exist?	Reference: *Physical Security Assessment for the Department of Veterans Affairs Facilities*	
11.2	Have critical air systems been rebalanced? If so, when and how often?	Although the system may function, it must be tested periodically to ensure it is performing as designed. Balancing is also critical after initial construction to set equipment to proper performance per the design. Rebalancing may only occur during renovation. Reference: *CDC/NIOSH Pub 2002-139*	
11.3	Is air pressurization monitored regularly?	Some areas require positive or negative pressure to function properly. Pressurization is critical in a hazardous environment or emergency situation. Measuring pressure drop across filters is an indication when filters should be changed, but also may indicate that low pressures are developing downstream and could result in loss of expected protection. Reference: *CDC/NIOSH Pub 2002-139*	
11.4	Does the building have a policy or procedure for periodic recommissioning of major Mechanical/Electrical/Plumbing (M/E/P) systems?	Recommissioning involves testing and balancing of systems to ascertain their capability to perform as described. Reference: *Physical Security Assessment for the Department of Veterans Affairs Facilities*	
11.5	Is there an adequate O&M program, including training of facilities management staff?	If O&M of critical systems is done with in-house personnel, management must know what needs to be done and the workforce must have the necessary training to ensure systems reliability. Reference: *CDC/NIOSH Pub 2002-139*	
11.6	What maintenance and service agreements exist for M/E/P systems?	When an in-house facility maintenance work force does not exist or does not have the capability to perform the work, maintenance and service contracts are the alternative to ensure critical systems will work under all	

ASSET VALUE, THREAT/HAZARD, VULNERABILITY, AND RISK

Building Vulnerability Assessment Checklist (Continued)

Section	Vulnerability Question	Guidance	Observations
		conditions. The facility management staff requires the same knowledge to oversee these contracts as if the work was being done by in-house personnel. Reference: *Physical Security Assessment for the Department of Veterans Affairs Facilities*	
11.7	Are backup power systems periodically tested under load?	Loading should be at or above maximum connected load to ensure available capacity and automatic sensors should be tested at least once per year. Periodically (once a year as a minimum) check the duration of capacity of backup systems by running them for the expected emergency duration or estimating operational duration through fuel consumption, water consumption, or voltage loss. Reference: *FEMA 386-7*	
11.8	Is stairway and exit sign lighting operational?	The maintenance program for stairway and exit sign lighting (all egress lighting) should ensure functioning under normal and emergency power conditions. Expect building codes to be updated as emergency egress lighting is moved from upper walls and over doorways to floor level as heat and smoke drive occupants to crawl along the floor to get out of the building. Signs and lights mounted high have limited or no benefit when obscured. Reference: *FEMA 386-7*	
12	**Security Systems**		
	Perimeter Systems		
12.1	Are black/white or color CCTV (closed circuit television) cameras used? Are they monitored and recorded 24 hours/7 days a week? By whom? Are they analog or digital by design? What are the number of fixed, wireless, and pan-tilt-zoom cameras used?	Security technology is frequently considered to complement or supplement security personnel forces and to provide a wider area of coverage. Typically, these physical security elements provide the first line of defense in deterring, detecting, and responding to threats and reducing vulnerabilities. They must be viewed as an integral component of the overall security program. Their design, engineering, installation, operation, and management must be able to meet daily security challenges from a cost-effective and efficiency perspective. During and after an incident, the system, or its backups, should be functional per the planned design. Consider color CCTV cameras to view and record activity at the perimeter of the building, particularly at primary entrances and exits. A mix of monochrome cameras	

ASSET VALUE, THREAT/HAZARD, VULNERABILITY, AND RISK

Building Vulnerability Assessment Checklist (Continued)

Section	Vulnerability Question	Guidance	Observations
	Who are the manufacturers of the CCTV cameras? What is the age of the CCTV cameras in use?	should be considered for areas that lack adequate illumination for color cameras. Reference: *GSA PBS P-100*	
12.2	Are the cameras programmed to respond automatically to perimeter building alarm events? Do they have built-in video motion capabilities?	The efficiency of monitoring multiple screens decreases as the number of screens increases. Tying the alarm system or motion sensors to a CCTV camera and a monitoring screen improves the man-machine interface by drawing attention to a specific screen and its associated camera. Adjustment may be required after installation due to initial false alarms, usually caused by wind or small animals. Reference: *Physical Security Assessment for the Department of Veterans Affairs Facilities*	
12.3	What type of camera housings are used and are they environmental in design to protect against exposure to heat and cold weather elements?	Reference: *Physical Security Assessment for the Department of Veterans Affairs Facilities*	
12.4	Are panic/duress alarm buttons or sensors used, where are they located, and are they hardwired or portable?	Call buttons should be provided at key public contact areas and as needed in offices of managers and directors, in garages and parking lots, and other high-risk locations by assessment. Reference: *GSA PBS P-100*	
12.5	Are intercom call boxes used in parking areas or along the building perimeter?	See Item 12.4.	
12.6	What is the transmission media used to transmit camera video signals: fiber, wire line, telephone wire, coaxial, wireless?	Reference: *Physical Security Assessment for the Department of Veterans Affairs Facilities*	
12.7	Who monitors the CCTV system?	Reference: *DOC CIAO Vulnerability Assessment Framework 1.1*	

ASSET VALUE, THREAT/HAZARD, VULNERABILITY, AND RISK

Building Vulnerability Assessment Checklist (Continued)

Section	Vulnerability Question	Guidance	Observations
12.8	What is the quality of video images both during the day and hours of darkness? Are infrared camera illuminators used?	Reference: *Physical Security Assessment for the Department of Veterans Affairs Facilities*	
12.9	Are the perimeter cameras supported by an uninterruptible power supply, battery, or building emergency power?	Reference: *Physical Security Assessment for the Department of Veterans Affairs Facilities*	
12.10	What type of exterior Intrusion Detection System (IDS) sensors are used? (electromagnetic; fiber optic; active infrared; bistatic microwave; seismic; photoelectric; ground; fence; glass break (vibration/shock); single, double, and roll-up door magnetic contacts or switches)	Consider balanced magnetic contact switch sets for all exterior doors, including overhead/roll-up doors, and review roof intrusion detection. Consider glass break sensors for windows up to scalable heights. Reference: *GSA PBS-P100*	
12.11	Is a global positioning system (GPS) used to monitor vehicles and asset movements?	Reference: *Physical Security Assessment for the Department of Veterans Affairs Facilities*	
	Interior Security		
12.12	Are black/white or color CCTV cameras used? Are they monitored and recorded 24 hours/7 days a week? By whom? Are they analog or digital by design? What are the number of fixed, wireless, and pan-tilt-zoom cameras used? Who are the manufacturers of the CCTV cameras? What is the age of the CCTV cameras in use?	See Item 12.1. Reference: *Physical Security Assessment for the Department of Veterans Affairs Facilities*	

ASSET VALUE, THREAT/HAZARD, VULNERABILITY, AND RISK

Building Vulnerability Assessment Checklist (Continued)

Section	Vulnerability Question	Guidance	Observations
12.13	Are the cameras programmed to respond automatically to interior building alarm events? Do they have built-in video motion capabilities?	The efficiency of monitoring multiple screens decreases as the number of screens increases. Tying the alarm system or motion sensors to a CCTV camera and a monitoring screen improves the man-machine interface by drawing attention to a specific screen and its associated camera. Reference: *Physical Security Assessment for the Department of Veterans Affairs Facilities*	
12.14	What type of camera housings are used and are they designed to protect against exposure or tampering?	Reference: *Physical Security Assessment for the Department of Veterans Affairs Facilities*	
12.15	Do the camera lenses used have the proper specifications, especially distance viewing and clarity?	Reference: *Physical Security Assessment for the Department of Veterans Affairs Facilities*	
12.16	What is the transmission media used to transmit camera video signals: fiber, wire line, telephone wire, coaxial, wireless?	Reference: *Physical Security Assessment for the Department of Veterans Affairs Facilities*	
12.17	Are the interior camera video images of good visual and recording quality?	Reference: *Physical Security Assessment for the Department of Veterans Affairs Facilities*	
12.18	Are the interior cameras supported by an uninterruptible power supply source, battery, or building emergency power?	Reference: *Physical Security Assessment for the Department of Veterans Affairs Facilities*	
12.19	What are the first costs and maintenance costs associated with the interior cameras?	Reference: *Physical Security Assessment for the Department of Veterans Affairs Facilities*	
12.20	What type of security access control system is used?	Reference: *Physical Security Assessment for the Department of Veterans Affairs Facilities*	

Building Vulnerability Assessment Checklist (Continued)

Section	Vulnerability Question	Guidance	Observations
	Are the devices used for physical security also used (integrated) with security computer networks (e.g., in place of or in combination with user ID and system passwords)?		
12.21	What type of access control transmission media is used to transmit access control system signals (same as defined for CCTV cameras)?	Reference: *Physical Security Assessment for the Department of Veterans Affairs Facilities*	
12.22	What is the backup power supply source for the access control systems? (battery, uninterruptible power supply)	Reference: *Physical Security Assessment for the Department of Veterans Affairs Facilities*	
12.23	What access control system equipment is used? How old are the systems and what are the related first and maintenance service costs?	Reference: *Physical Security Assessment for the Department of Veterans Affairs Facilities*	
12.24	Are panic/duress alarm sensors used? Where are they located? Are they hardwired or portable?	Call buttons should be provided at key public contact areas and as needed in offices of managers and directors, in garages and parking lots, and other high-risk locations by assessment. Reference: *GSA PBS P-100*	
12.25	Are intercom call-boxes or a building intercom system used throughout the building?	See Item 12.24.	
12.26	Are magnetometers (metal detectors) and x-ray equipment used? At what locations within the building?	Reference: *DOC CIAO Vulnerability Assessment Framework 1.1*	

ASSET VALUE, THREAT/HAZARD, VULNERABILITY, AND RISK

Building Vulnerability Assessment Checklist (Continued)

Section	Vulnerability Question	Guidance	Observations
12.27	What type of interior IDS sensors are used: electromagnetic; fiber optic; active infrared-motion detector; photoelectric; glass break (vibration/shock); single, double, and roll-up door magnetic contacts or switches?	Consider magnetic reed switches for interior doors and openings. Reference: *GSA PBS-P100*	
12.28	Are mechanical, electrical, gas, power supply, radiological material storage, voice/data telecommunication system nodes, security system panels, elevator and critical system panels, and other sensitive rooms continuously locked, under electronic security, CCTV camera, and intrusion alarm systems surveillance?	Reference: *DOC CIAO Vulnerability Assessment Framework 1.1*	
12.29	What types of locking hardware are used throughout the building? Are manual and electromagnetic cipher, keypad, pushbutton, panic bar, door strikes, and related hardware and software used?	As a minimum, electric utility closets, mechanical rooms, and telephone closets should be secured. The mailroom should also be secured, allowing only authorized personnel into the area where mail is screened and sorted. Separate the public access area from the screening area for the postulated mailroom threats. All security locking arrangements on doors used for egress must comply with *NFPA 101, Life Safety Code*. Reference: *GSA PBS-P100*	
12.30	Are any potentially hazardous chemicals, combustible, or toxic materials stored on site in non-secure and non-monitored areas?	The storage, use, and handling locations should also be kept away from other activities. The concern is that an intruder need not bring the material into the building if it is already there and accessible. Reference: *Physical Security Assessment for the Department of Veterans Affairs Facilities*	
12.31	What security controls are in place to handle the processing of mail and protect against potential biological, explosive, or other threatening exposures?	Reference: *Physical Security Assessment for the Department of Veterans Affairs Facilities*	

ASSET VALUE, THREAT/HAZARD, VULNERABILITY, AND RISK

Building Vulnerability Assessment Checklist (Continued)

Section	Vulnerability Question	Guidance	Observations
12.32	Is there a designated security control room and console in place to monitor security, fire alarm, and other building systems? Is there a backup control center designated and equipped? Is there off-site 24-hour monitoring of intrusion detection systems?	Monitoring can be done at an off-site facility, at an on-site monitoring center during normal duty hours, or at a 24-hour on-site monitoring center. Reference: *GSA PBS-P100*	
12.33	Is the security console and control room adequate in size and does it provide room for expansion? Does it have adequate environment controls (e.g., a/c, lighting, heating, air circulation, backup power)? Is it ergonomically designed?	Reference: *Physical Security Assessment for the Department of Veterans Affairs Facilities*	
12.34	Is the location of the security room in a secure area with limited, controlled, and restricted access controls in place?	Reference: *Physical Security Assessment for the Department of Veterans Affairs Facilities*	
12.35	What are the means by which facility and security personnel can communicate with one another (e.g., portable radio, pager, cell phone, personal data assistants (PDAs))? What problems have been experienced with these and other electronic security systems?	Reference: *Physical Security Assessment for the Department of Veterans Affairs Facilities*	
12.36	Is there a computerized security incident reporting system used to prepare reports and track security incident trends and patterns?	Reference: *Physical Security Assessment for the Department of Veterans Affairs Facilities*	
12.37	Does the current security force have access to a computerized guard tour system?	This system allows for the systematic performance of guard patrols with validation indicators built in. The system notes stations/locations checked or missed, dates	

ASSET VALUE, THREAT/HAZARD, VULNERABILITY, AND RISK

Building Vulnerability Assessment Checklist (Continued)

Section	Vulnerability Question	Guidance	Observations
		and times of such patrols, and who conducted them on what shifts. Management reports can be produced for recordkeeping and manpower analysis purposes. Reference: *Physical Security Assessment for the Department of Veterans Affairs Facilities*	
12.38	Are vaults or safes in the building? Where are they located?	Basic structural design requires an understanding of where heavy concentrations of floor loading may occur so as to strengthen the floor and structural framing to handle this downward load. Security design also needs this information to analyze how this concentrated load affects upward and downward loadings under blast conditions and its impact upon progressive collapse. Location is important because safes can be moved by blast so that they should be located away from people and away from exterior windows. Vaults, on the other hand, require construction above the building requirements with thick masonry walls and steel reinforcement. A vault can provide protection in many instances due to its robust construction. Safes and vaults may also require security sensors and equipment, depending upon the level of protection and defensive layers needed. Reference: *U.S. Army TM 5-85*	
	Security System Documents		
12.39	Have security system as-built drawings been generated and are they ready for review?	Drawings are critical to the consideration and operation of security technologies, including its overall design and engineering processes. These historical reference documents outline system specifications and layout security devices used, as well as their application, location, and connectivity. They are a critical resource tool for troubleshooting system problems, and replacing and adding other security system hardware and software products. Such documents are an integral component to new and retrofit construction projects. Reference: *Physical Security Assessment for the Department of Veterans Affairs Facilities*	
12.40	Have security system design and drawing standards been developed?	Reference: *Physical Security Assessment for the Department of Veterans Affairs Facilities*	
12.41	Are security equipment selection criteria defined?	Reference: *Physical Security Assessment for the Department of Veterans Affairs Facilities*	

Building Vulnerability Assessment Checklist (Continued)

Section	Vulnerability Question	Guidance	Observations
12.42	What contingency plans have been developed or are in place to deal with security control center redundancy and backup operations?	Reference: *Physical Security Assessment for the Department of Veterans Affairs Facilities*	
12.43	Have security system construction specification documents been prepared and standardized?	Reference: *Physical Security Assessment for the Department of Veterans Affairs Facilities*	
12.44	Do all security system documents include current as-built drawings?	Reference: *Physical Security Assessment for the Department of Veterans Affairs Facilities*	
12.45	Have qualifications been determined for security consultants, system designers/engineers, installation vendors, and contractors?	Reference: *Physical Security Assessment for the Department of Veterans Affairs Facilities*	
12.46	Are security systems decentralized, centralized, or integrated? Do they operate over an existing IT network or are they a standalone method of operation?	Reference: *Physical Security Assessment for the Department of Veterans Affairs Facilities*	
12.47	What security systems manuals are available?	Reference: *Physical Security Assessment for the Department of Veterans Affairs Facilities*	
12.48	What maintenance or service agreements exist for security systems?	Reference: *Physical Security Assessment for the Department of Veterans Affairs Facilities*	

Building Vulnerability Assessment Checklist (Continued)

Section	Vulnerability Question	Guidance	Observations
13	**Security Master Plan**		
13.1	Does a written security plan exist for this site or building? When was the initial security plan written and last revised? Who is responsible for preparing and reviewing the security plan?	The development and implementation of a security master plan provides a roadmap that outlines the strategic direction and vision, operational, managerial, and technological mission, goals, and objectives of the organization's security program. Reference: *DOC CIAO Vulnerability Assessment Framework 1.1*	
13.2	Has the security plan been communicated and disseminated to key management personnel and departments?	The security plan should be part of the building design so that the construction or renovation of the structure integrates with the security procedures to be used during daily operations. Reference: *Physical Security Assessment for the Department of Veterans Affairs Facilities*	
13.3	Has the security plan been benchmarked or compared against related organizations and operational entities?	Reference: *Physical Security Assessment for the Department of Veterans Affairs Facilities*	
13.4	Has the security plan ever been tested and evaluated from a benefit/cost and operational efficiency and effectiveness perspective?	Reference: *Physical Security Assessment for the Department of Veterans Affairs Facilities*	
13.5	Does the security plan define mission, vision, and short- and long- term security program goals and objectives?	Reference: *Physical Security Assessment for the Department of Veterans Affairs Facilities*	
13.6	Are threats/hazards, vulnerabilities, and risks adequately defined and security countermeasures addressed and prioritized relevant to their criticality and probability of occurrence?	Reference: *DOC CIAO Vulnerability Assessment Framework 1.1*	
13.7	Has a security implementation schedule been established to address recommended security solutions?	Reference: *Physical Security Assessment for the Department of Veterans Affairs Facilities*	

ASSET VALUE, THREAT/HAZARD, VULNERABILITY, AND RISK

Building Vulnerability Assessment Checklist (Continued)

Section	Vulnerability Question	Guidance	Observations
13.8	Have security operating and capital budgets been addressed, approved, and established to support the plan?	Reference: *Physical Security Assessment for the Department of Veterans Affairs Facilities*	
13.9	What regulatory or industry guidelines/standards were followed in the preparation of the security plan?	Reference: *Physical Security Assessment for the Department of Veterans Affairs Facilities*	
13.10	Does the security plan address existing security conditions from an administrative, operational, managerial, and technical security systems perspective?	Reference: *Physical Security Assessment for the Department of Veterans Affairs Facilities*	
13.11	Does the security plan address the protection of people, property, assets, and information?	Reference: *Physical Security Assessment for the Department of Veterans Affairs Facilities*	
13.12	Does the security plan address the following major components: access control, surveillance, response, building hardening, and protection against CBR and cyber-network attacks?	Reference: *Physical Security Assessment for the Department of Veterans Affairs Facilities*	
13.13	Has the level of risk been identified and communicated in the security plan through the performance of a physical security assessment?	Reference: *Physical Security Assessment for the Department of Veterans Affairs Facilities*	
13.14	When was the last security assessment performed? Who performed the security risk assessment?	Reference: *DOC CIAO Vulnerability Assessment Framework 1.1*	

ASSET VALUE, THREAT/HAZARD, VULNERABILITY, AND RISK

Building Vulnerability Assessment Checklist (Continued)

Section	Vulnerability Question	Guidance	Observations
13.15	**Are the following areas of security analysis addressed in the security master plan?** **Asset Analysis:** Does the security plan identify and prioritize the assets to be protected in accordance to their location, control, current value, and replacement value? **Threat Analysis:** Does the security plan address potential threats; causes of potential harm in the form of death, injury, destruction, disclosure, interruption of operations, or denial of services? (possible criminal acts [documented and review of police/security incident reports] associated with forced entry, bombs, ballistic assault, biochemical and related terrorist tactics, attacks against utility systems infrastructure and buildings) **Vulnerability Analysis:** Does the security plan address other areas associated with the site or building and its operations that can be taken advantage of to carry out a threat? (architectural design and construction of new and existing buildings, technological support systems [e.g., heating, air conditioning, power, lighting and security systems, etc.] and operational procedures, policies, and controls) **Risk Analysis:** Does the security plan address the findings from the asset, threat/hazard, and vulnerability analyses in order to develop, recommend, and consider implementation of appropriate security countermeasures?	This process is the input to the building design and what mitigation measures will be included in the facility project to reduce risk and increase safety of the building and people. Reference: *USA TM 5-853, Security Engineering*	

ASSET VALUE, THREAT/HAZARD, VULNERABILITY, AND RISK

Appendix III

Glossary of General Terms

A

Access Control: Any combination of barriers, gates, electronic security equipment, and/or guards that can deny entry to unauthorized personnel or vehicles.

Access Control Point: A station at an entrance to a building where identification is checked and people and hand-carried items are searched.

Access Controls: Procedures and controls that limit or detect access to minimum essential infrastructure resource element such as people, technology, applications, data and/or facilities, thereby protecting these resources against loss of integrity, confidentiality, accountability and/or availability.

Access Control System: Also referred to as an electronic entry control system; it is an electronic system that controls building entry and egress.

Access Control System Elements: Detection measures used to control vehicle or personnel entry into a protected area. Elements include locks, electronic entry control systems and guards.

Access Group: A software configuration of an access control system that group together access points or authorized users for easier arrangement and system maintenance.

Access Road: Any roadway such as a maintenance, delivery, service, emergency, or other special limited use road that is necessary for the operation of a building or structure.

Accountability: The explicit assignment of responsibilities for oversight of areas of control to executives, managers, staff, owners, providers and users of minimum essential infrastructure elements.

Active Vehicle Barrier: An impediment placed at an access control point which may be manually or automatically deployed in response to detection of a threat.

Aerosol: A fine liquid or solid particles suspended in a gas; for example, fog or smoke.

Aggressor: Any person seeking to compromise a function or a structure.

Airborne Contamination: Chemical or biological agents introduced into and fouling the source of supply breathing or conditioning air.

Airlock: A building entry configuration with which airflow from the outside can be prevented from entering a toxic free area. An airlock used two doors, only one of which can be opened at a time, and a blower system to maintain positive air pressures and purge contaminated air from the airlock before the second door is opened.

Alarm Assessment: Verification and evaluation of an alarm alert, through the use of closed circuit television or human observation. Systems used for alarm assessment are designed to responds rapidly, automatically, and predictably to the receipt of alarms at the security center.

Alarm Printers: Alarm printers provide a hard copy of all alarm events and system activity, as well as limited backup in case the visual display fails.

Alarm Priority: A hierarchy of alarms by orders of importance. This is often used in larger systems to give priority to alarms with greater importance.

Annunciation: A visual, audible, or other indicator by a security system of a condition.

Anti-terrorism: Defensive measures used to reduce the vulnerability of individuals, forces, and property to terrorist acts.

Assessment: The evaluation and interpretation of measurements and other information to provide a basis for decision-making.

Asset: A resource of value that requires protection. An asset can be tangible (e.g., people, buildings, facilities, equipment, activities and operations, as well as information; or intangible, such as processes, or a company's information and reputation).

Asset Protection: Security programs designed to protect personnel, facilities, and equipment; accomplished through planned and integrated application of combating terrorism, physical security, operations security, personnel protective services.

Attack: A hostile action that results in the destruction, injury or death to people, or damage or destruction to public/private property.

Audible Alarm Devices: An alarm device that produces an audible announcement; e.g., bell, horn, siren, etc.

B

Ballistics Attack: Attack in which small arms—pistols, submachine guns, shotguns and rifles—are fired from a distance.

Biological Agents: Living organisms or the materials derived from them, that cause disease in, or harm to humans, animals, or plants.

Blast Curtains: Heavy curtains made of blast-resistant materials that could protect occupants from flying debris.

Blast Resistant Glazing: Window opening glazing that is resistant to blast effects.

Blast Vulnerability Envelope: The area in which an explosive device will cause damage to assets.

Bollard: A vehicle barrier consisting of a cylinder usually made of steel and filled with concrete. It is placed on end in the ground and spaced approximately three feet apart, preventing vehicles from passing but allowing pedestrian movement.

Boundary Penetration Sensors: Interior intrusion detection sensors which detect an attempt by individuals to penetrate or enter the building.

Building Hardening: Enhanced construction that reduces vulnerability to external blast and ballistic attack.

Building Separation: The distance between the closest points on exterior walls of adjacent buildings/structures.

C

Capacitance Sensor: A device that detects an intruder approaching or touching a metal object by sensing a change in capacitance between the object and the ground.

Card Reader: A device that gathers or reads information when a card is presented as identification, method.

Chemical Agent: A chemical substance that is intended to kill, seriously injure, or incapacitate people. Generally separated by severity of effect (lethal, blister, and incapacitating).

Clear Zone: An area that is clear of visual obstructions and landscape that could conceal either a threat or perpetrator.

Closed Circuit Television (CCTV): An electronic system of cameras, control equipment, recorders, and related equipment used for surveillance.

Chimney Effect: Air movement in a building between floors caused by differential air temperature between the air inside and outside the building. It occurs in vertical shafts, such as elevators, stairwells and a conduit/wiring/piping chase. Hotter air inside the building will rise and be replaced by infiltration with colder outside air through the lower portions of the building. Conversely, reversing the temperature will reverse the flow down the chimney. Also known as stack effect.

Collateral Damage: Injury or damage to assets that are not the primary target of an attack.

Components and Cladding: Elements of the building envelope that do not qualify as part of the main wind-force resisting system.

Control Center: A centrally located room or facility staffed by personnel charged with the over sight of specific situations and/or equipment.

Controlled Area: An area into which access is controlled or limited.

Controlled Lighting: Lighting illumination of specific areas or sections.

Controlled Perimeter: Physical boundary at which vehicle and personnel access is controlled at the perimeter of a site. Should demonstrate the capability to search vehicles and individuals.

Conventional Construction: Building construction that is not specifically designed to resist weapons, explosives, or chemical, biological and radiological effects. Designed to resist common loadings and environmental effects such as wind, seismic and snow loads only.

Covert Entry: Attempts to enter a facility using false credentials or stealth.

Crash Bar: A mechanical egress device located on the interior side of a door that unlocks the door when pressure is applied in the direction of egress.

Crisis Management: The measures taken to identify, acquire, and plan the use of resources needed to anticipate, prevent, and/or resolve a threat or act of terrorism.

Critical Assets: Those assets essential to the minimum operations of the organization, and to ensure the health and safety of the general public.

Critical Infrastructure: Primary infrastructure systems (utilities, telecommunications, transportation, etc.,) whose incapacity would have a debilitating impact on the organization's ability to function.

D

Damage Assessment: The process used to appraise or determine the number of injuries and deaths, damage to public and private property/facilities resulting from a natural or human-engineered disaster.

Decontamination: The reduction or removal of a chemical, biological or radiological material from the surface of a structure, area, object, or person.

Defense Layer: Building design or exterior perimeter barriers intended to delay attempted forced entry.

Delay Rating: A measure of the effectiveness of penetration protection of a defense layer.

Design Basis Threat: The threat—tactics, associated weapons, tools or explosives—against which assets within a building must be protected and upon which the security engineering design of the building is based.

Design Team: A group if individuals from various engineering and architectural disciplines responsible for the protective system design.

Detection Layer: A ring of intrusion detection sensors located on or adjacent to a defensive layer or between two defensive layers.

Detection Measures: Protective measures which detect intruders, weapons, or explosives; assist in assessing the validity of detection; control access to protected areas; and communicate the appropriate information to the response force. Detection measures include detection system, assessment system, and access control system elements.

Detection System Elements: Detection measures which detect the presence of intruders, weapons, or explosives. Detection system elements include intrusion detection systems, weapons and explosives devices, and guards.

Disaster: An occurrence of a natural catastrophe, technological accident or human-caused event that results in deaths, and/or multiple injuries, and severe property damage.

Domestic Terrorism: The unlawful use, or threatened use, of force or violence by a group or individual based and operating within the United States or Puerto Rico without foreign direction, committed against persons or property to intimidate or coerce a government, the civilian population, or any segment thereof in furtherance of political or social objectives.

Door Position Switch: A switch that changes state based on whether or not a door is closed. Typically, a switch mounted in a frame that is actuated by a magnet in a door.

Door Strike, Electronic: An electro-mechanical lock that releases a door plunger to unlock the door. Typically, an electronic door strike is mounted in place of or near a normal door strike plate.

Dose Rate (Radiation): A general term indicating the total quantity of ionizing radiation or energy absorbed by a person or animal per unit of time.

Dosimeter: An instrument for measuring and registering total accumulated exposure to ionizing radiation.

Duress Alarm Devices: Also know as panic buttons, these devices are designated specifically to initiate a panic alarm.

E

Electronic Entry Control Systems: Electronic devices which automatically verify authorization for persons to enter or exit a controlled area.

Electronic Security System: An integrated system which encompasses interior and exterior sensors, closed-circuit television systems for assessment of alarm conditions, electronic entry control systems, data transmission media, and alarm reporting systems for monitoring, control and display of various alarm and system information.

Emergency: Any natural or man-made occurrence that results in substantial injury or harm to the population or substantial damage to or loss of property.

Emergency Alert System: A communications system of broadcast stations and interconnecting facilities authorized by the Federal Communication Commission. The system provides the President of the United States and other national, state, and local officials the means to broadcast emergency information to the public before, during, and after disasters.

Emergency Operations Center: The protected site from which state and local civil government officials coordinate, monitor, and direct emergency response activities during an emergency.

Emergency Operations Plan: A document that describes how people and property will be protected during a disaster and disaster threat situations. It details who is responsible for carrying out specific actions; identities key personnel, equipment, facilities, supplies and other resources available for use during the disaster. It also outlines how actions will be coordinated.

Emergency Public Information: Information which is disseminated primarily in anticipation of an emergency or at the actual time of an emergency. Additionally, it provides information, directs actions, instructs and transmits direct orders.

Entry-Control Stations: These stations should be provided at main perimeter entrances where security personnel are present. Entry-control stations should be located as close as possible to the perimeter entrance, thereby permitting personnel inside the station to maintain constant surveillance over the entrance and its approaches.

Evacuation: Organized, phased and supervised dispersal of people from dangerous or potentially dangerous areas.

Evacuees: All persons removed or moving from areas threatened or struck by a disaster.

Explosives Disposal Container: A small container into which small quantities of explosives may be placed to contain their blast pressures and fragments if the explosive detonates.

F

Fence Protection: An intrusion detection technology that detects a person crossing a fence by various methods such as climbing, crawling, cutting, etc.

Fence Sensors: Exterior intrusion detection sensors which detect aggressors as they attempt to climb over, cut through, or otherwise disturb a fence.

Fiber Optics: A method of data transfer by passing bursts of light through a strand of glass or clear plastic.

First Responder: Local police, fire and emergency medical personnel who first arrive on the scene of an incident and who take action to save lives, protect property and meet basic human needs.

Forced Entry: Entry to a denied area achieved through force to create an opening in fence, walls, doors, etc., or to overpower guards and security personnel.

Fragment Retention Film: Thin, optically clear film applied to glass to minimize the spread of glass fragments when the glass is shattered.

G

Glare Security-Lighting: Illumination projected from a secure perimeter into the surrounding area making it possible to see potential intruders at a considerable distance while making it difficult to observe activities within the secure perimeter.

Glass-Break Detector: Intrusion detection sensors that are designed to detect breaking glass either through vibration or acoustics.

Glazing: A material installed in a sash, ventilator, or panes such as glass, plastic, etc., including material such as thin granite installed in a curtain wall.

Grid Wire Sensors: Intrusion detection sensors that use a grid of wires to cover a wall or fence. An alarm is sounded if the wires are cut.

H

Hazard: A source of potential danger or an adverse condition.

Hazard Mitigation: An action that is taken to reduce or eliminate the long-term risk to human life and property from hazards. It is also used to define the cost-effective measures that reduce the potential for damage to a facility from a disaster occurrence.

Hazardous Material: Any substance that poses a risk to people's health, safety and/or property. These substances include explosives, radioactive materials, flammable liquids or solids, combustible liquids or solids, poisons, oxidizers, toxins and corrosive materials.

High-Hazard Areas: Geographic locations that have been determined, through historical experience and vulnerability analysis, to be likely to experience the effects of a specific hazard (either natural or human-engineered) that results in significant property damage and loss of life.

High-Risk Target: Any resource or facility that, because of mission sensitivity, ease of access, isolation, and symbolic value, may be an especially attractive or accessible terrorist target.

Human-Engineered Hazards: Human-engineered hazards are both technological hazards and terrorism. They originate from human activity.

I

Impact Analysis: A management level analysis that identifies the impact of losing an entity's resources. It measures the effect of resource loss and escalating losses over time in order to provide the entity with reliable data upon which to base decisions on hazard mitigation and business continuity planning.

Insider Compromise: A person authorized access to a facility (an insider) compromises assets by taking advantage of that authorized accessibility.

International Terrorism: Violent acts or acts dangerous to human life that violate the criminal laws of the United States or any state; or that would be a criminal violation if committed within the jurisdiction of the United States, or any state. Acts of international terrorism appear to be intended to intimidate or coerce a civilian population, influence the policy of a government by intimidation or coercion, or affect the conduct of a government by assassination or kidnapping. International terrorist acts occur outside the United States, or transcend national boundaries in terms of the means by which they are accomplished; the persons targeted for coercion or intimidation; or the areas/geographical locale in which the perpetrators operate, or seek asylum.

Intrusion Detection Sensors: Devices that initiate alarm signals by sensing the stimulus, change, or condition for which they were designed.

Intrusion Detection System: The combination of components, including sensors, control units, transmission lines, and monitor units that are integrated to operate in a specified manner.

Isolated Fenced Perimeters: Fenced perimeters with 100 feet or more space outside the fence that is clear of obstruction, thus making approach obvious.

J

Jersey Barrier: A protective concrete barrier used as a highway divider. It is also used as a method for traffic speed control at entrance gates and to keep vehicles away from buildings.

L

Laminated Glass: A flat like of uniform thickness consisting of two monolithic glass plies bonded together with an interlayer material. Many different interlayer materials are used in laminated glass.

Landscaping: The use of plantings with or without landforms and/or large boulders to act as a perimeter barrier against defined threats.

Laser Card: A card technology that uses a laser reflected off of a card for the unique identification of the card.

Layers of Protection: A traditional approach in security engineering using concentric circles extending out from an area to be protected as demarcation points for different security strategies.

Level of Protection: The degree to which an asset is protected against injury or damage from an attack.

Liaison: An official sent to facilitate communications and coordination.

Limited Area: A restricted area within close proximity of a security interest. Uncontrolled movement may permit access to the item. Escorts and other internal restrictions may prevent access to the item.

Line of Sight: Direct observation between two points with the naked eye or hand-held optics.

Line-of-Sight Sensor: A pair of devices used as an intrusion detection sensor that monitor any movement through the field between the sensors.

Line Supervision: A data integrity strategy that monitors the communications link for connectivity and tampering. In intrusion detection system sensors, line supervision is often referred to as two-state, three-state, or four-state in respect to the number of conditions monitored.

M

Magnetic Lock: An electro-magnetic lock that unlocks a door when the power is removed.

Magnetic Stripe: A card technology that uses a magnetic stripe on the card to encode data used for unique identification of the card.

Mail-Bomb Delivery: Bombs or incendiary devices delivered to the target in letters or packages.

Man-Trap: An access control strategy that uses a pair of interlocking doors to prevent tailgating. Only one door can be unlocked at a time.

Mass Notification: Capability to provide real-time information to all building occupants/personnel in the immediate vicinity of the facility during emergency situations.

Microwave Motion Sensors: Intrusion detection sensors that use microwave energy to sense movement with the sensors field of view.

Minimum Measures: Protective measures that can be applied to all buildings regardless of the identified threat. These measures offer defense or detection opportunities for minimal cost, facilitate future upgrades, and may deter acts of aggression.

Mitigation: The actions taken to reduce the exposure to and the impact of an attack or disaster.

Motion Detector: An intrusion detection sensor that changes state, based on movement in the sensors' field of view.

Moving Vehicle Bomb: An explosive-laden car or truck driven into or near a building and detonated.

N

Non-Exclusive Zone: An area around an asset that has controlled entry but shared or less restrictive access than an exclusive zone.

Non-Persistent Agent: An agent that, upon release, loses its ability to cause casualties after 10 to 15 minutes. It has a high evaporation rate, is lighter than air, and will disperse rapidly. A short-term hazard but will be more persistent in small, unventilated areas.

Nuclear, Biological or Chemical Weapons: Also called Weapons of Mass Destruction (WMD). Weapons that are characterized by their capability to produce mass casualties.

Nuclear Detonation: An explosion resulting from fission and/or fusion reactions in nuclear material, such as that from a nuclear weapon.

O

Open System Architecture: An IT term that systems are capable of interfacing with other systems from any vendor, which also uses open system architecture. The opposite would be a proprietary system.

Organizational Areas of Control: Controls consist of the policies, procedures, practices and organization structures designed to provide reasonable assurance that business objectives will be achieved and that undesired events will be prevented or detected and corrected.

P

Passive Infrared Motion Sensors: Devices that detect a change in the thermal energy pattern caused by a moving intruder and initiate an alarm when the change in energy satisfies the detector's alarm criteria.

Passive Vehicle Barrier: A vehicle barrier which is permanently deployed and does not require response to be effective.

Perimeter Barrier: A fence, wall, vehicle barrier, landform, or line of vegetation applied along an exterior perimeter used to obscure vision, hinder personnel access, or hinder or prevent vehicle access.

Persistent Agent: An agent when released retains its casualty producing effects for an extended period of time—from 30 minutes to several days. A persistent agent usually has a low evaporation rate and its vapor is heavier than air. Consequently, its vapor cloud tends to hug the ground. It is also considered a long-term hazard. Along with inhalation hazards, skin contact should also be avoided.

Physical Security: The part of security concerned with measures and concepts designed to safeguard personnel, prevent unauthorized access to equipment, installations, material and documents. Physical security also safeguards the aforementioned against espionage, sabotage, damage and theft.

Planter Barrier: A passive vehicle barrier that is usually constructed of concrete, and filled with dirt. Planters and bollards are the items used to keep vehicles away from existing buildings. The size and depth of installation below grade is the planter's determining factor in stopping any vehicle from entering the area.

Plume: Airborne material that spreads from a particular source and the dispersal of particles, gases, vapors and aerosols into the atmosphere.

Polycarbonate Glazing: A plastic glazing material with enhanced resistance to ballistics or blast effects.

Preparedness: Establishing the plans, training, exercises and resources necessary to enhance mitigation of and achieve readiness for and recovery from all hazards, disaster and emergencies including weapons of mass destruction incidents.

Pressure Mat: A mat that generates an alarm when pressure is applied to any part of the mat's surface. Pressure mats can be used to detect an intruder approaching a protected object or act as detection sensors when placed by doors and windows.

Primary Asset: An asset which is the ultimate target by a terrorist.

Primary Gathering Building: Inhabited buildings that are occupied by 50 or more personnel.

Probability of Detection: A measure of an intrusion detection sensor's performance in detecting an intruder within its detection zone.

Probability of Intercept: The probability that an act of aggression will be detected and that a response force will intercept the aggressor before the asset can be compromised.

Progressive Collapse: A chain reaction failure with damage that may result to the upper floors of a building collapsing onto the lower floors.

Protective Barriers: Those barriers that define the physical limits of a site or area. They restrict, channel or impede access and form a continuous obstacle around the object.

Protective Measures: The elements of a protective system that protect an asset against a threat. Protective measures include defensive and detection measures.

Protective System: An integration of all the protective measures required to protect an asset against the range of threats applicable to the asset.

Proximity Sensors: Intrusion detection sensors that change state based on the contact or proximity of a human to the sensor. These sensors often measure the change in capacitance as a human body enters the measured field.

Public Information Officer: A federal, state, or local government official responsible for preparing and coordinating the dissemination of emergency public information.

R

Radiation: High energy particles or gamma rays that are emitted by an atom as the substance undergoes radioactive decay. Particles can be either charged alpha or beta particles or neutral neutron or gamma rays.

Radiation Sickness: The symptoms characterizing the sickness known as radiation injury, resulting from excessive exposure of the whole body to ionizing radiation.

Radiological Monitoring: The process of locating and measuring radiation by means of survey instruments that can detect and measure ionizing radiation.

Recovery: The long-term activities beyond the initial crisis period and emergency response phase of disaster operations that focus on returning all systems in the community to a normal status or to reconstitute the systems to a new condition that is less vulnerable.

Request-to-Exit Device: Passive infrared motion sensors that are used to signal an electronic entry system that egress is imminent or to unlock a door.

Resource Management: Those actions taken by a government to identify sources and obtain resources needed to support response activities; coordinate the supply, allocation, distribution, and delivery of resources so that they arrive where and when they are most needed.

Response: The plan of action implemented to perform those duties and services intended to preserve and protect life and property, and to provide services to the surviving population.

Response Time: The length of time from the instant an attack is detected to the instant a security force arrives at the scene.

Restricted Area: The area with access controls that is subject to special restrictions, or controls for security reasons.

RF Data Transmission: A communications link using radio frequency to send or receive data.

Risk: The potential for loss of, or damage to an asset. It is measured based upon the value of the asset in relation to the threats and vulnerabilities associated with it.

S

Safe Haven: Secure areas within the interior of the facility. A safe haven should be designed so that it requires more time to penetrate by terrorists than it takes for the response force to reach the protected area. A safe area protected from physical attack, or air-isolated from CBR contamination.

Secure/Access Mode: The state of an area monitored by an intrusion detection system in regards to how alarm conditions are reported.

Security Analysis: The method of studying the nature of and the relationship between assets, threats and vulnerabilities.

Security Console: Specialized furniture, racking and related apparatus used to house the security equipment required in a control center.

Shielded Wire: Wire with a conductive wrap used to mitigate electromagnetic currents.

Situational Crime Prevention: A crime prevention strategy based on reducing the opportunities for crime by increasing the effort required to commit a crime, increasing the risks associated with committing the crime, and reducing the target appeal or vulnerability (i.e., person or property). This opportunity reduction is achieved by management and uses policies such as training and procedures, as well as physical approaches such as alternation of the built environment.

Smart Card: A card technology that allows data to be written, stored, and read on a card that is used for identification and/or access.

Specific Threat: Known or postulated terrorist activity that is focused on targeting a particular asset.

Standoff Distance: A distance maintained between a building/portion of the building and the potential location for an explosive detonation, or other threat.

Stationary Vehicle Bomb: An explosive-laden car/truck that is stopped or parked near a building.

Structural Protective Barriers: Man-made devices (e.g., fences, walls, floors, roofs, grills, bars, signs, etc.) used to restrict, or impede access.

Superstructure: The supporting elements of a building above the foundation.

Supplies-Bomb Delivery: Bombs or incendiary devices that are concealed and delivered to supply or material handling points such as loading docks.

T

Tactics: The specific methods of achieving the perpetrator's goals to injure personnel, destroy assets, or steal material/information.

Tamper Switch: An intrusion detection sensor that monitors and equipment enclosure for breach.

Technological Hazard: Any incident that can arise from human activities, such as the manufacture, transportation, storage and used of hazardous materials. Such hazards are normally assumed to be accidental and their consequences unintended.

Terrorism: The unlawful use of force and violence against persons or property to intimidate or coerce a government, the civilian population, or any segment thereof, in furtherance of political or social objectives.

Thermally Tempered Glass: Heat-treated glass that has a higher tensile strength and resistance to blast pressures, although a greater susceptibility to airborne debris.

Threat: Any indication, circumstance, or event with the potential to cause loss of, or damage to an asset.

Threat Analysis: A continual process of compiling and examining all available information concerning potential threats and human-engineered hazards. A common method to evaluate terrorist groups is to review the factors of existence, capability, intentions, history and targeting.

Time/Date Stamp: Date inserted into a CCTV video signal with the time and date of the video as it was created.

Toxicity: A measure of the harmful effects produced by a given amount of a toxin on a living organism.

Toxic-Free Area: An area within a facility in which the air supply is free of toxic chemical or biological agents.

U

Unobstructed Space: Space around an inhabited building without obstruction large enough to conceal explosive devise 150 mm (6 inches) or greater in height.

V

Vault: A reinforced room for securing items.

Vertical Rod: Typical door hardware often used with a crash bar to lock a door by inserting rods vertically from the door into the door frame.

Vibration Sensors: Intrusion detection sensor that change state when vibration is present.

Video Intercom System: An intercom system that also incorporates a small CCTV system for verification.

Video Motion Detection: Motion detection technology that looks for changes in the pixels of a video image.

Video Displays: A display or monitor used to inform the operator visually of the status of the electronic security system.

Visual Surveillance: The aggressor uses ocular and photographic devices including binoculars and cameras with telephoto lenses. This equipment monitors facility, installation operations, or to monitor assets.

Voice Recognition: A biometric technology based on nuances of the human voice.

Volumetric Motion Sensors: Interior intrusion detection sensors which are designed to sense aggressor motion within a protected space.

Vulnerability: Any weakness in an asset or mitigation measure that can be exploited by an aggressor, adversary or competitor. Vulnerability refers to the organization's susceptibility to injury.

W

Warning: The alerting of emergency response personnel and the public to the threat of extraordinary danger and the related effects that specific hazards may cause.

Waterborne Contamination: Chemical, biological, or radiological agents introduced into a water supply.

Weapons-Grade Material: Nuclear material considered most suitable for a nuclear weapon. It usually refers to uranium enriched to above 90% uranium-235 or plutonium with greater than approximately 90% plutonium-239.

Weapons of Mass Destruction: Any explosive, incendiary, or poison gas, bomb, grenade, rocket having a propellant charge of more than 4 ounces; or a missile having an explosive incendiary charge of more than 0.25 ounce, or a mine or device similar to the above. It also refers to poison gas, weapons involving a disease organism, or weapons that are designed to release radiation or radioactivity at a level that is dangerous to human life. It is intended to cause death or serious injury to persons, or significant damage to property.

Glossary of Chemical Terms

Acetyl cholinesterase: An enzyme that hydrolyzes the neurotransmitter acetylcholine. The action of this enzyme is inhibited by nerve agents.

Aerosol: Fine liquid or solid particles suspended in a gas; for example, fog or smoke.

Atropine: A compound used as an antidote for nerve agents.

Casualty (toxic) agents: Produce incapacitation, serious injury, or death. They can be used to incapacitate or kill victims. These agents are the choking, blister, nerve and blood agents.

    Blister agents: Substances that cause blistering of the skin. Exposure is through liquid or vapor contact with any exposed tissue (eyes, skin, and lungs). Examples include distilled mustard, nitrogen mustard, and lewisite.

    Blood agents: Substances that injure a person by interfering with cell respiration (the exchange of oxygen and carbon dioxide between blood and tissues). Examples are arsine (SA), cyanogens chloride (CK), hydrogen chloride, and hydrogen cyanide (AC).

    Choking/lung/pulmonary agents: Substances that cause physical injury to the lungs. Exposure is through inhalation. In extreme cases, membranes swell and lungs become filled with liquid. Death results from lack of oxygen; hence, the victim is "choked." Examples include chlorine, diphosgene, cyanide, nitrogen oxide, and zinc oxide.

    Nerve agents: Substances that interfere with the central nervous system. Exposure is primarily through contact with the liquid— skin and eyes—and secondarily through inhalation of the vapor. Three distinct symptoms associated with nerve agents are: pinpoint pupils, an extreme headache, and severe tightness in the chest. Also see the G-series and V-series nerve agents.

Chemical agents: A chemical substance that is intended for use in military operations to kill, seriously injure, or incapacitate people through its physiological effects. Excluded from consideration are riot control agents, and smoke and flame materials. The agent may appear as a vapor, aerosol, or liquid; it can be either a casualty/toxic agent or an incapacitating agent.

Cutaneous: Pertaining to the skin.

Decontamination: The process of making any person, object, or area safe by absorbing, destroying, neutralizing, making harmless, or removing the hazardous material.

G-series nerve agents: Chemical agents of moderate to high toxicity developed in the 1930's. Examples include Tabun, Sarin, Soman, Phosphonofluoridic acid, ethyl- and cyclohexyl sarin.

Incapacitating agents: Produce temporary physiological and/or mental effects via action on the central nervous system. Effects may persist for hours or days, but victims usually do not require medical treatments. However, such treatment speeds recovery.

> **Vomiting agents**: Produce nausea and vomiting effects, can also cause coughing, sneezing, pain in the nose and throat, nasal discharge, and tears.
>
> **Tear/riot control-agents**: Produce irritating or disabling effects that rapidly disappear within minutes after exposure ceases. Examples include chloroacetophenone (known commercially as Mace), chloropicrin, tear gas and Capsaicin (pepper spray).
>
> **Central nervous system depressants**: Compounds that have the predominant effect of depressing or blocking the activity of the central nervous system. The primary mental effects include the disruption of the ability to think, sedation, and lack of motivation.
>
> **Central nervous system stimulants**: Compounds that have the predominant effect of flooding the brain with too much information. The primary mental effect is loss of concentration, causing indecisiveness and the inability to act in a sustained, purposeful manner.

Industrial agents: Chemicals developed or manufactured for use in industrial operations or research by industry, government, or academia. These chemicals are not primarily manufactured for the specific purpose of producing human casualties or rendering equipment, facilities, or areas dangerous for use by man. Hydrogen cyanide, cyanogen chloride, phosgene, chloropicrin and many herbicides and pesticides are industrial chemicals that also can be chemical agents.

Liquid agent: A chemical agent that appears to be an oily film or droplets. The color ranges from clear to brownish color.

Nonpersistent agent: An agent that upon release loses its ability to cause casualties after 10-15 minutes. It has a high evaporation rate, is lighter than air and will disperse rapidly. It is considered to be a short-term hazard. However, in small unventilated areas, the agent will be more persistent.

Organophosphorous compound: A compound, containing the elements phosphorus and carbon, whose physiological effects include inhibition of acetyl cholinesterase. Many pesticides—malathione and parathion—and virtually all nerve agents are organophosphorous compounds.

Percutaneous agent: Able to be absorbed by the body through the skin.

Persistent agent: An agent that upon release retains its casualty-producing effects for an extended period of time, usually anywhere from 30 minutes to several days. A persistent agent usually has a low evaporation rate and its vapor is heavier than air. Therefore, its vapor cloud tends to hug the ground. It is considered to be a long-term hazard. Although inhalation hazards are still a concern, extreme caution should be taken to avoid skin contact as well.

Protection: Any means by which an individual protects his body. Measures include masks, self-contained breathing apparatuses, clothing, structures such as buildings, and vehicles.

V-series nerve agents: Chemical agents of moderate to high toxicity developed in the 1950's. They are generally persistent.

Vapor agent: A gaseous form of a chemical agent. If heavier than air, the cloud will be close to the ground. If lighter than air, the cloud will rise and disperse more quickly.

Volatility: A measure of how readily a substance will vaporize.

Glossary of Biological Terms

Aerosol: Fine liquid or solid particles suspended in a gas; for example, fog or smoke.

Antibiotic: A substance that inhibits the growth of or kills microorganisms.

Antisera: The liquid part of blood containing antibodies that react against disease causing agents such as those used in biological warfare.

Bacteria: Single-celled organisms that multiply by cell division which can cause disease in humans, plants, or animals.

Biochemicals: The chemicals that make up or are produced by living things.

Biological warfare agents: Living organisms or the materials derived from them that cause disease in or harm to humans, animals, or plants, or cause deterioration of material. Biological agents may be used as liquid droplets, aerosols, or dry powders.

Biological warfare: The intentional use of biological agents as weapons to kill or injure humans, animals, or plants, or to damage equipment.

Bioregulators: Biochemicals that regulate bodily functions. Bioregulators that are produced by the body are termed "endogenous." Some of these same bioregulators can be chemically synthesized.

Causative agent: The organism or toxin that is responsible for causing a specific disease or harmful effect.

Contagious: Capable of being transmitted from one person to another.

Culture: A population of micro-organisms grown in a medium.

Decontamination: The process of making people, objects, or areas safe by absorbing, destroying, neutralizing, making harmless, or removing the hazardous material.

Fungi: Any of a group of plants mainly characterized by the absence of chlorophyll, the green colored compound found in other plants. Fungi range from microscopic single-celled plants—such as molds and mildews—to large plants—such as mushrooms.

Host: An animal or plant that harbors or nourishes another organism.

Incapacitating agent: Agents that produce physical or psychological effects, or both, that may persist for hours or days after exposure, rendering victims incapable of performing normal physical and mental tasks.

Infectious agents: Biological agents capable of causing disease in a susceptible host.

Infectivity: 1. The ability of an organism to spread. 2. The number of organisms required to cause an infection to secondary hosts. 3. The capability of an organism to spread out from the site of infections and cause disease in the host organism. Infectivity also can be viewed as the number of organisms required to cause an infection.

Line-source delivery system: A delivery system in which the biological agent is dispersed from a moving ground or air vehicle in a line perpendicular to the direction of the prevailing wind.

Mycotoxin: A toxin produced by fungi.

Microorganism: Any organism, such as bacteria, viruses, and some fungi, that can be seen only with a microscope.

Nebulizer: A device for producing a fine spray or aerosol.

Organism: Any individual living thing, whether animal or plant.

Parasite: Any organism that lives in or on another organism without providing benefit in return.

Pathogen: Any organism—usually living—capable of producing serious disease or death, such as bacteria, fungi, and viruses.

Pathogenic agents: Biological agents capable of causing serious disease.

Point-source delivery system: A delivery system in which the biological agent is dispersed from a stationary position. This delivery method results in coverage over a smaller area than with the line-source system.

Route of exposure—entry: The path by which a person comes into contact with an agent or organism; for example, through breathing, digestion, or skin contact.

Single-cell protein: Protein-rich material obtained from cultured algae, fungi, protein and bacteria, and often used as food or animal feed.

Spore: A reproductive form some micro-organisms can take to become resistant to environmental conditions, such as extreme heat or cold, while in a "resting stage."

Toxicity: A measure of the harmful effect produced by a given amount of a toxin on a living organism. The relative toxicity of an agent can be expressed in milligrams of toxin needed per kilogram of body weight to kill experimental animals.

Toxins: Poisonous substances produced by living organisms.

Vaccine: A preparation of killed or weakened microorganism products used to artificially induce immunity against a disease.

Vector: An agent, such as an insect or rat, capable or transferring a pathogen from one organism to another.

Venom: A poison produced in the glands of some animals; for example, snakes, scorpions, or bees.

Virus: An infectious micro-organism that exists as a particle rather than as a complete cell. Particle sizes range from 20 to 400 manometers—one-billionth of a meter. Viruses are not capable or reproducing outside of a host cell.

Glossary of Radiological Terms

Acute radiation syndrome: Consists of three levels of effects: Hematopoletic—blood cells, most sensitive; Gastrointestinal—GI cells, very sensitive; and Central Nervous System—brain/muscle cells, insensitive. The initial signs and symptoms are nausea, vomiting, fatigue, and loss of appetite. Below about 200 rems, these symptoms may be the only indication of radiation exposure.

Alpha particle: The alpha particle has a very short range in air and a very low ability to penetrate other materials, but it has a strong ability to ionize materials. Alpha particles are unable to penetrate even the thin layer of dead cells of human skin and consequently are not an external radiation hazard. Alpha-emitting nuclides inside the body as a result of inhalation or ingestion are a considerable internal radiation hazard.

Beta particles: High-energy electrons emitted from the nucleus of an atom during radioactive decay. They normally can be stopped by the skin or a very thin sheet of metal.

Cesium-137 (Cs-137): A strong gamma ray source and can contaminate property, entailing extensive clean-up. It is commonly used in industrial measurement gauges and for irradiation of material. Half-life is 30.2 years.

Cobalt-60 (Co-60): A strong gamma ray source. It is extensively used as a radiotherapeutic for treating cancer, food and material irradiation, gamma radiography, and industrial measurement gauges. Half-life is 5.27 years.

Curie (Ci): A unit of radioactive decay rate defined as 3.7×10^{10} disintegrations per second.

Decay: The process by which an unstable element is changed to another isotope or another element by the spontaneous emission of radiation from its nucleus. This process can be measured by using radiation detectors such as Geiger counters.

Decontamination: The process of making people, objects, or areas safe by absorbing, destroying, neutralizing, making harmless, or removing the hazardous material.

Dose: A general term for the amount of radiation absorbed over a period of time.

Dosimeter: A portable instrument for measuring and registering the total accumulated dose to ionizing radiation.

Gamma rays: High-energy photons emitted from the nucleus of atoms; similar to x rays. They can penetrate deeply into body tissue and many materials. Cobalt-60 and Cesium-137 are both strong gamma ray emitters. Shielding against gamma radiation requires thick layers of dense materials, such as lead. Gamma rays are potentially lethal to humans.

Half-life: The amount of time needed for half of the atoms of a radioactive material to decay.

Highly enriched uranium (HEU): Uranium that is enriched to above 20% Uranium-235 (U-235). Weapons-grade HEU is enriched to above 90% in U-235.

Ionize: To split off one or more electrons from an atom, thus leaving it with a positive electric charge. The electrons usually attach to one of the atoms or molecules, giving them a negative charge.

Iridium-192: A gamma-ray emitting radioisotope used for gamma-radiography. The half-life is 73.83 days.

Isotope: A specific element always has the same number of protons in the nucleus. That same element may, however, appear in forms that have different numbers of neutrons in the nucleus. These different forms are referred to as "isotopes" of the element. For example, deuterium (2H) and tritium (3H) are isotopes of ordinary hydrogen (H).

Lethal dose (50/30): The dose of radiation expected to cause death within 30 days to 50% of those exposed without medical treatment. The generally accepted range from 400-500 rem received over a short period of time.

Nuclear reactor: A device in which a controlled, self-sustaining nuclear chain reaction can be maintained with the use of cooling to remove generated heat.

Plutonium-239 (PU-239): A metallic element used for nuclear weapons. The half-life is 24,110 years.

Radiation: High energy alpha or beta particles of gamma rays that are emitted by an atom as the substance undergoes radioactive decay.

Radiation sickness: Symptoms resulting from excessive exposure to radiation of the body.

Radioactive waste: Disposable, radioactive materials resulting from nuclear operations. Wastes are generally classified into two categories, high-level and low-level waste.

Radiological Dispersal Device (RDD): A device—weapon or equipment—

other than a nuclear explosive device, designed to disseminate radioactive material in order to cause destruction, damage, or injury by means of the radiation produced by the decay of such material.

Radioluminescence: The luminescence produced by particles emitted during radioactive decay.

REM: A Roentgen Man Equivalent is a unit of absorbed dose that takes into account the relative effectives of radiation that harms human health.

Shielding: Materials—lead, concrete, etc.—used to block or attenuate radiation for protection of equipment, materials, or people.

Special Nuclear Material (SNM): Plutonium and uranium enriched in the isotope Uranium-233 or Uranium 235.

Uranium 235 (U-235): Naturally occurring uranium U-235 is found at 0.72% enrichment. U-235 is used as a reactor fuel or for weapons; however, weapons typically use U-235 enriched to 90%.

X-ray: An invisible, highly penetrating electromagnetic radiation of much shorter wavelength (higher frequency) than a visible light; very similar to gamma-rays.

Website Directory
Associations & Organizations

American Lifelines Alliance
http://www.americanlifelinesalliance.org

Applied Technology Council
http://www.atcouncil.org

Center for Strategic and International Studies
http://www.csis.org

Centers for Disease Control and Prevention / National Institute for
Occupational Safety and Health
http://www.cdc.gov/niosh

Central Intelligence Agency
http://www.cia.gov

Federal Aviation Administration
http://www.faa.gov

Institute of Transportation Engineers
http://www.ite.org

International CPTED (Crime Prevention Through Environmental
Design) Association
http://www.cpted.net.home.amt

Lawrence Berkeley National Laboratory
http://www.securebuildings.lbl.gov

National Academy of Science
http://www4.nationalacademies.org/nas/nashome.nsf

National Defense Industrial Association
http://www.ndia.org

Security Design Coalition
http://www.designingforsecurity.org

Security Industry Association
http://www.siaonline.org/

Technical Support Working Group (Departments of Defense and State)
http://www.www.tswg.gov

U.S. Air Force Electronic System Center, Hanscom Air Force Base
http://www.eschq.hanscom.af.mil/

U.S. Army Soldiers and Biological Chemical Command: Basic
Information on Building Protection
http://www.buildingprotection.sbccom.army.mil

U.S. Department of Justice
http://www.usdoj.gov

The Infrastructure Security Partnership
http://www.tsip.org

American Council of Engineering Companies
http://www.acec.org

The American Institute of Architects, Security Resource Center
http://www.aia.org/security

American Society of Civil Engineers
http://www.asce.org

Associated General Contractors of America
http://www.agc.org

Construction Industry Institute
http://construction-institute.org

Federal Emergency Management Agency
http://www.fema.gov

National Institute of Standards and Technology, Building and Fire
Research Laboratory
http://www.bfrl.nist.gov

Naval Facilities Engineering Command
http://www.navfac.navy.mil

Society of American Military Engineers
http://www.same.org

U.S. Army Corp of Engineers
http://www.usace.army.mil

Air Conditioning Contractors of America
http://www.acca.org

Air-Conditioning and Refrigeration Institute, Inc.
http://www.ari.org

Airport Consultants Council
http://www.acconline.org

Alliance for Fire & Smoke Containment & Control
http://www.afscconline.org

American Association of State Highway and Transportation
http://www.transportation.org

American Institute of Chemical Engineers, Center for Chemical Process
Safety
http://www.aiche.org/ccps

American Planning Association
http://www.planning.org

American Portland Cement Alliance
http://www.portcement.org/apca

American Public Works Association
http://www.apwa.net

American Railway Engineering & Maintenance of Way Association
http://www.arema.org

American Society for Industrial Security International
http://www.asisonline.org

American Society of Heating, Refrigerating, and Air Conditioning
Engineers
http://www.ashrae.org

American Society of Interior Designers
http://www.asid.org

American Society of Mechanical Engineers
http://www.asme.org

American Underground Construction Association
http://www.auca.org
http://www.auaonline.org

American Water Resources Association
http://www.awra.org

Associated Locksmiths of America
http://www.aloa.org

Association of Metropolitan Water Agencies
http://www.amwa.net

Association of State Dam Safety Officials
http://www.damsafety.org

Building Futures Council
http://www.thebfc.com

Building Owners and Manager Association International, Emergency Resource Center
http://www.boma.org/emergency

California Department of Health Services, Division of Drinking Water & Environmental Management
http://www.dhs.cahwnet.gov/ps/ddwen

Construction Industry Roundtable
http://www.cirt.org

Construction Innovation Forum
http://www.cif.org

Construction Specifications Institute
http://www.scinet.org

Construction Users Roundtable
http://www.curt.org

Defense Threat Reduction Agency
http://www.dtra.mil

Design-Build Institute of America
http://www.dbia.org

Drexel (University) Intelligent Infrastructure & Transportation Safety Institute
http://www.di3.drexel.edu

Federal Highway Administration
http://www.fhwa.dot.gov

Florida Department of Transportation, Emergency Management Office
http://www.11.myflorida.com/safety/Emp/emp.htm

Florida Department of Community Affairs, Division of Emergency Management
http://www.floridadisaster.org/bpr/EMTOOLS/Severe/terrorism.htm
http://www.dca.state.fl.us/bpr/EMTOOLS/CIP/critical-infrasturcture-protecti.
    htm

George Washington University, Institute for Crisis, Disaster, and Risk
Management
http://www.cee.seas.gwu.edu
http://www.seas.gwu.edu/~icdm

Homeland Protection Institute, Ltd.
http://www.hpi-tech.org

Inland Rivers Port and Terminals
http://www.irpt.net

Institute of Electrical and Electronics Engineers, Inc.-USA
http://www.ieeeusa.org
http://www.ieee.or/portal/index.jsp

International Association of Foundation Drilling
http://www.adsc-iafd.com

International Code Council
http://www.intlcode.org

International Facility Management Association
http://www.ifma.org

Market Development Alliance of the FRP Composites Industry
http://www.mdacomposite.org

Multidisciplinary Center for Earthquake Engineering Research
http://www.mceer.buffalo.edu

National Aeronautics and Space Administration
http://www.nasa.gov

National Capital Planning Commission
http://www.ncpc.gov

National Center for Manufacturing Sciences
http://www.ncms.org

National Concrete Masonry Association
http://www.ncma.org

National Conference of States on Building Codes and Standards
http://www.ncsbcs.org

National Council of Structural Engineers Associations
http://www.ncsea.com
http://dwp.bigplanet.com/engineers/homepage

National Crime Prevention Institute
http://www.louisville.edu/a-s/ja/ncpi/courses.htm

National Fire Protection Institute
http://www.nfpa.org

National Institute of Building Sciences
http://www.nibs.org
http://www.wbdg.org

National Park Service, Denver Service Center
http://www.nps.gov/dsc

National Precast Concrete Association
http://www.precast.org

National Wilderness Training Center, Inc.
http://www.wildernesstraining.net

New York City Office of Emergency Preparedness
http://www.nyc.gov/html/oem

Ohio State University
http://www.osu.edu/homelandsecurity

Pentagon Renovation Program
http://www.renovation.pentagon.mil

Portland Cement Association
http://www.portcement.org

Primary Glass Manufacturers Council
http://www.primaryglass.org

Protective Glazing Council
http://www.protectiveglazing.org

Protective Technology Center at Penn State University
http://www.ptc.psu.edu

SAVE International
http://www.value-eng.org

Society of Fire Protection Engineers
http://www.sfpe.org

Southern Building Code Congress, International
http://www.sbcci.org

Sustainable Buildings Industry Council
http://www.sbicouncil.org

Transit Standards Consortium
http://www.tsconsortium.org

Transportation Research Board / Marine Board
http://www.trb.org

Transportation Security Administration – Maritime and Land
http://www.tsa.dot.gov

U.S. Air Force Civil Engineer Support Agency
http://www.afcesa.af.mil

U.S. Coast Guard
http://www.uscg.mil

U.S. Department of Energy
http://www.energy.gov

U.S. Department of Health and Human Services
http://www.hhs.gov

U.S. Department of Veteran Affairs
http://www.va.gov/fascmgt

U.S. Environmental Protection Agency, Chemical Emergency Preparedness and Prevention Office – Counter-terrorism
http://www.epa.gov/swercepp/cntr-ter.html

U.S. General Services Administration
http://www.gsa.gov

U.S. Green Building Council
http://www.usgbc.org

U.S. Marine Corps Headquarters
http://www.usmc.mil

U.S. Society on Dams
http://www.ussdams.org

University of Missouri, Department of Civil and Environmental Engineering, National Center for Explosion Resistant Design
http://www.engineering.missouri.edu/explosion.htm

Virginia Polytechnic Institute and State University
http://www.ce.vt.edu

Water and Wastewater Equipment Manufacturers Association
http://www.wwema.org

The Partnership for Critical Infrastructure
http://www.pcis.org

Department of Commerce Critical Infrastructure Assurance Office
http://www.ciao.gov

Department of Energy
http://www.energy.gov

Department of Homeland Security
http://www.whitehouse.gov/deptofhomeland

National Infrastructure Protection Center
http://www.nipc.gov

Anser Institute for Homeland Security
http://www.homelandsecurity.org

The Financial Services Roundtable Technology Group
http://www.bitsinfo.org

CERT® Coordination Center
http://www.cert.org

Electronic Warfare Associates
http://www.ewa.com

Information Technology Association of America
http://www.itaa.org

The Institute for Internal Auditors
http://www.theiia.org

National Cyber Security Alliance
http://www.staysafeonline.infor

North American Electric Reliability Council
http://www.nerc.com

SANS Institute
http://www.sans.org

The U.S. Chamber of Commerce, Center for Corporate Citizenship
http://www.uschamber.com/ccc

Association of Metropolitan Water Agencies
http://www.amwa.net

The Council of State Governments
http://www.csg.org

International Association of Emergency Managers
http://www.iaem.com

National Association of State CIOs
http://www.nascio.org

National Emergency Managers Association
http://www.nemaweb.org

National Governor's Association
http://www.nga.org

The National League of Cities
http://www.nlc.org

Index